T0076140

ENERGY SCIENCE, ENGINEERING AND TECHNOLOGY

CARBON CAPTURE AND SEQUESTRATION

DEVELOPMENT EFFORTS AND OUTLOOK

ENERGY SCIENCE, ENGINEERING AND TECHNOLOGY

Additional books in this series can be found on Nova's website under the Series tab.

Additional E-books in this series can be found on Nova's website under the E-book tab.

CLIMATE CHANGE AND ITS CAUSES, EFFECTS AND PREDICTION

Additional books in this series can be found on Nova's website under the Series tab.

Additional E-books in this series can be found on Nova's website under the E-book tab.

ENERGY SCIENCE, ENGINEERING AND TECHNOLOGY

CARBON CAPTURE AND SEQUESTRATION

DEVELOPMENT EFFORTS AND OUTLOOK

ANDERSON CARTER MITCHELL
AND
ROSS FREEMAN
EDITORS

New York

Copyright © 2013 by Nova Science Publishers, Inc.

All rights reserved. No part of this book may be reproduced, stored in a retrieval system or transmitted in any form or by any means: electronic, electrostatic, magnetic, tape, mechanical photocopying, recording or otherwise without the written permission of the Publisher.

For permission to use material from this book please contact us:
Telephone 631-231-7269; Fax 631-231-8175
Web Site: http://www.novapublishers.com

NOTICE TO THE READER

The Publisher has taken reasonable care in the preparation of this book, but makes no expressed or implied warranty of any kind and assumes no responsibility for any errors or omissions. No liability is assumed for incidental or consequential damages in connection with or arising out of information contained in this book. The Publisher shall not be liable for any special, consequential, or exemplary damages resulting, in whole or in part, from the readers' use of, or reliance upon, this material. Any parts of this book based on government reports are so indicated and copyright is claimed for those parts to the extent applicable to compilations of such works.

Independent verification should be sought for any data, advice or recommendations contained in this book. In addition, no responsibility is assumed by the publisher for any injury and/or damage to persons or property arising from any methods, products, instructions, ideas or otherwise contained in this publication.

This publication is designed to provide accurate and authoritative information with regard to the subject matter covered herein. It is sold with the clear understanding that the Publisher is not engaged in rendering legal or any other professional services. If legal or any other expert assistance is required, the services of a competent person should be sought. FROM A DECLARATION OF PARTICIPANTS JOINTLY ADOPTED BY A COMMITTEE OF THE AMERICAN BAR ASSOCIATION AND A COMMITTEE OF PUBLISHERS.

Additional color graphics may be available in the e-book version of this book.

Library of Congress Cataloging-in-Publication Data

ISBN: 978-1-62257-810-8

Published by Nova Science Publishers, Inc. ✝ *New York*

CONTENTS

PREFACE

Carbon capture and sequestration (or storage), known as CCS, has attracted congressional interest as a measure for mitigating global climate change because large amounts of carbon dioxide (CO_2) emitted from fossil fuel use in the United States are potentially available to be captured and stored underground and prevented from reaching the atmosphere. Large, industrial sources of CO_2, such as electricity-generating plants, are likely candidates for CCS because they are predominantly stationary, single-point sources. Electricity generation contributes over 40% of U.S. CO_2 emissions from fossil fuels. Currently, U.S. power plants do not capture large volumes of CO_2 for CCS. This book provides an overview of what CCS is, how it is supposed to work, why it has gained the interest and support of some members of Congress, and what some of the challenges are to its implementation and deployment across the United States.

Chapter 1 - Carbon capture and sequestration (or storage)—known as CCS—has attracted congressional interest as a measure for mitigating global climate change because large amounts of carbon dioxide (CO_2) emitted from fossil fuel use in the United States are potentially available to be captured and stored underground and prevented from reaching the atmosphere.

Large, industrial sources of CO_2, such as electricity-generating plants, are likely initial candidates for CCS because they are predominantly stationary, single-point sources.

Electricity generation contributes over 40% of U.S. CO_2 emissions from fossil fuels. Currently, U.S. power plants do not capture large volumes of CO_2 for CCS.

Several projects in the United States and abroad—typically associated with oil and gas production—are successfully capturing, injecting, and storing

CO_2 underground, albeit at relatively small scales. The oil and gas industry in the United States injects nearly 50 million tons of CO_2 underground each year for the purpose of enhanced oil recovery (EOR). The volume of CO_2 envisioned for CCS as a climate mitigation option is overwhelming compared to the amount of CO_2 used for EOR.

According to the U.S. Department of Energy (DOE), the United States has the potential to store billions of tons of CO_2 underground and keep the gas trapped there indefinitely. Capturing and storing the equivalent of decades or even centuries of CO_2 emissions from power plants (at current levels of emissions) suggests that CCS has the potential to reduce U.S. greenhouse gas emissions substantially while allowing the continued use of fossil fuels.

An integrated CCS system would include three main steps: (1) capturing and separating CO_2 from other gases; (2) purifying, compressing, and transporting the captured CO_2 to the sequestration site; and (3) injecting the CO_2 in subsurface geological reservoirs or storing it in the oceans. Deploying CCS technology on a commercial scale would be a vast undertaking.

The CCS process, although simple in concept, would require significant investments of capital and of time.

Capital investment would be required for the technology to capture CO_2 and for the pipeline network to transport the captured CO_2 to the disposal site.

Time would be required to assess the potential CO_2 storage reservoir, inject the captured CO_2, and monitor the injected plume to ensure against leaks to the atmosphere or to underground sources of drinking water, potentially for years or decades until injection activities cease and the injected plume stabilizes.

Three main types of geological formations in the United States are being considered for storing large amounts of CO_2: oil and gas reservoirs, deep saline reservoirs, and unmineable coal seams. The deep ocean also has a huge potential to store carbon; however, direct injection of CO_2 into the deep ocean is controversial, and environmental concerns have forestalled planned experiments in the open ocean. Mineral carbonation—reacting minerals with a stream of concentrated CO_2 to form a solid carbonate—is well understood, but it is still an experimental process for storing large quantities of CO_2.

Large-scale CCS injection experiments are only beginning in the United States to test how different types of reservoirs perform during CO_2 injection of a million tons of CO_2 or more. Results from the experiments will undoubtedly be crucial to future permitting and site approval regulations.

Acceptance by the general public of large-scale deployment of CCS may be a significant challenge. Some of the large-scale injection tests could garner information about public acceptance, as citizens become familiar with the concept, process, and results of CO_2 injection tests in their local communities.

Chapter 2 – On March 27, 2012, the U.S. Environmental Protection Agency (EPA) proposed a new rule that would limit emissions to no more than 1,000 pounds of carbon dioxide (CO_2) per megawatt-hour of production from new fossil-fuel power plants with a capacity of 25 megawatts or larger.

EPA proposed the rule under Section 111 of the Clean Air Act. According to EPA, new natural gas-fired combined-cycle power plants should be able to meet the proposed standards without additional cost. However, new coal-fired plants would only be able to meet the standards by installing carbon capture and sequestration (CCS) technology.

The proposed rule has sparked increased scrutiny of the future of CCS as a viable technology for reducing CO_2 emissions from coal-fired power plants.

The proposed rule also places a new focus on whether the U.S. Department of Energy's (DOE's) CCS research, development, and demonstration (RD&D) program will achieve its vision of developing an advanced CCS technology portfolio ready by 2020 for large-scale CCS deployment.

Congress has appropriated nearly $6 billion since FY2008 for CCS RD&D at DOE's Office of Fossil Energy: approximately $2.3 billion from annual appropriations and $3.4 billion from the American Recovery and Reinvestment Act (or Recovery Act).

The large and rapid influx of funding for industrial-scale CCS projects from the Recovery Act may accelerate development and deployment of CCS in the United States.

However, the future deployment of CCS may take a different course if the major components of the DOE program follow a path similar to DOE's flagship CCS demonstration project, FutureGen, which has experienced delays and multiple changes of scope and design since its inception in 2003. A question for Congress is whether FutureGen represents a unique case of a first mover in a complex, expensive, and technically challenging endeavor, or whether it indicates the likely path for all large CCS demonstration projects once they move past the planning stage.

Since enactment of the Recovery Act, DOE has shifted its RD&D emphasis to the demonstration phase of carbon capture technology. The shift appears to heed recommendations from many experts who called for large, industrial-scale carbon capture demonstration projects (e.g., 1 million tons of CO_2 captured per year). Funding from the Recovery Act for large-scale demonstration projects was 40% of the total amount of DOE funding for all CCS RD&D from FY2008 through FY2012.

To date, there are no commercial ventures in the United States that capture, transport, and inject industrial-scale quantities of CO2 solely for the purposes of carbon sequestration. However, CCS RD&D in 2012 is just now embarking on commercial-scale demonstration projects for CO2 capture, injection, and storage. The success of these projects will likely bear heavily on the future outlook for widespread deployment of CCS technologies as a strategy for preventing large quantities of CO_2 from reaching the atmosphere while U.S. power plants continue to burn fossil fuels, mainly coal.

Given the pending EPA rule, congressional interest in the future of coal as a domestic energy source appears directly linked to the future of CCS. In the short term, congressional support for building new coal-fired power plants could be expressed through legislative action to modify or block the proposed EPA rule. Alternatively, congressional oversight of the CCS RD&D program could help inform decisions about the level of support for the program and help Congress gauge whether it is on track to meet its goals.

Chapter 3 - Electricity generation in the United States depends heavily on the use of coal: Coal-fired power plants produce 40 percent to 45 percent of the nation's electricity. At the same time, those facilities account for roughly a third of all U.S. emissions of carbon dioxide (CO_2), which together with other greenhouse gases has become increasingly concentrated in the atmosphere. Most climate scientists believe that the buildup of those gases could have costly consequences.

One much-discussed option for reducing the nation's greenhouse gas emissions while preserving its ability to produce electricity at coal-fired power plants is to capture the CO_2 that is emitted when the coal is burned, compress it into a fluid, and then store it deep underground. That process is commonly called carbon capture and storage (CCS). Although the process is in use in some industries, no CCS-equipped coal-fired power plants have been built on a commercial scale because any electricity generated by such plants would be much more expensive than electricity produced by conventional coal-burning plants. Utilities, rather than federal agencies, make most of the decisions about investments in the electricity industry, and today they have little incentive to equip their facilities with CCS technology to lessen their CO_2 emissions.

Since 2005, lawmakers have provided the Department of Energy (DOE) with about $6.9 billion to further develop CCS technology, demonstrate its commercial feasibility, and reduce the cost of electricity generated by CCS-equipped plants. But unless DOE's funding is substantially increased or other policies are adopted to encourage utilities to invest in CCS, federal support is likely to play only a minor role in deployment of the technology.

Engineers have estimated that, on average, electricity generated by the first CCS-equipped commercial-scale plants would initially be about 75 percent more costly than electricity generated by conventional coal-fired plants. (Most of that additional cost is attributable to the extra facilities and energy that would be needed to capture the CO_2.) That initial cost differential would probably shrink, however, as the technology became more widely applied and equipment manufacturers and construction companies became more familiar with it— a pattern of cost reduction called learning-by-doing.

DOE aims to bring down the additional costs for generating electricity with CCS technology to no more than 35 percent, or less than half the current cost premium. Such a cost differential, if combined with a tax on carbon or policies restricting CO_2 emissions, could allow coal-fired plants with CCS to be competitive with those without CCS.

Such a reduction in costs might be accomplished over time through learning-by-doing, which would require that a certain amount of new generating capacity be built—in the form of new coal-fired CCS-equipped generating plants. Using the historical pace of reductions in costs for earlier emission-control technologies, the Congressional Budget Office (CBO) estimates that more than 200 gigawatts (GW) of coal-fired generating capacity with CCS capabilities will have to be built to meet DOE's cost reduction goal. That estimate of new capacity, which is equivalent to about two-thirds of the total current capacity of U.S. coal-powered electricity generation plants, is subject to considerable uncertainty. Nevertheless, in the absence of a significant technological breakthrough, it seems clear that a large amount of new CCS capacity—installed either at new plants or, through retrofitting, at existing plants—would be needed to reduce costs by enough to achieve DOE's goal.

But the demand for electricity in the United States is growing slowly, and even if DOE's cost reduction target was attained, coal-fired power plants equipped with CCS technology would not be competitive with coal-fired plants that lacked it unless policies restricting CO2 emissions or imposing a price on them were adopted. Consequently, under current laws and policies, utilities are unlikely to build that much new generating capacity—that is, more than 200 GW—or invest in adding CCS technology to much of their existing capacity for many decades. If, however, new policies restricted or imposed a price on CO_2 emissions, the domestic stock of electricity generation plants would turn over more rapidly, and CCS technology would become more competitive economically, increasing the potential for construction of CCS-equipped plants in the United States. Nevertheless, investors already have several options for generating electricity—nuclear power, wind, biomass, other renewables, and natural gas—that produce few, if any, CO_2 emissions. The amount of investment in CCS would depend on how costs for the different alternatives compared with costs for electricity generation without CCS.

Reductions in costs for CCS-equipped power plants could also come from experience outside the United States. Demand for electricity is growing rapidly in other parts of the world—for example, China and India—and those countries are increasing their capacity to satisfy it. If plants that were equipped with early versions of the CCS technology were built abroad or if some coal-fired power plants now in operation in other countries were retrofitted with CCS, the cost of generating electricity at plants that were subsequently built or retrofitted in the United States would be expected to be lower than the cost of generating electricity at the plants that were built initially. At present, however, foreign investment in CCS, like investment in the technology in the United States, centers not on building full-scale CCS-equipped commercial plants but on conducting research and development, carrying out small-scale demonstrations of the technology's feasibility, and building pilot plants.

Until now, most efforts to develop CCS have focused on coal-fired power plants. However, the price of natural gas has dropped substantially in recent years, and the share of electricity generated by natural gas-fired plants has expanded and is likely to continue to grow. The cost of producing electricity with a natural gas-fired plant equipped with CCS could be lower, depending on future prices for coal and natural gas, than the cost of producing electricity with a coal-fired CCS plant. At present, though, regulatory action to curb CO_2 emissions is more likely to shift electricity production from coal to natural gas (without CCS) and other low-emission fuels, such as biomass, rather than to CCS-capable plants.

CBO's analysis suggests that unless the federal government adopts policies that encourage or require utilities to generate electricity with fewer greenhouse gas emissions, the projected high cost of using CCS technology means that DOE's current program for developing CCS is unlikely to do much to support widespread use of the technology. A number of other policy approaches could be considered. For example, lawmakers could redirect resources that now fund technology demonstration projects toward research and development, for which the rationale for federal involvement is strongest and the record of success better. Alternatively, policymakers could impose costs on users of electricity whose generation releases greenhouse gases—for example, through a tax on carbon—thereby making CCS more competitive, or they could experiment with different types of electricity production subsidies that would provide more incentive for private-sector investments in CCS. As another option, lawmakers could reduce or eliminate future spending for CCS, leaving most of the potential for further development of CCS technology to countries with high rates of growth in the demand for electricity and in the need for new electricity-generating capacity.

In: Carbon Capture and Sequestration   ISBN: 978-1-62257-810-8
Editors: A.C. Mitchell and R. Freeman   © 2013 Nova Science Publishers, Inc.

Chapter 1

CARBON CAPTURE AND SEQUESTRATION (CCS): A PRIMER[*]

Peter Folger

SUMMARY

Carbon capture and sequestration (or storage)—known as CCS—has attracted congressional interest as a measure for mitigating global climate change because large amounts of carbon dioxide (CO_2) emitted from fossil fuel use in the United States are potentially available to be captured and stored underground and prevented from reaching the atmosphere.

Large, industrial sources of CO_2, such as electricity-generating plants, are likely initial candidates for CCS because they are predominantly stationary, single-point sources.

Electricity generation contributes over 40% of U.S. CO_2 emissions from fossil fuels. Currently, U.S. power plants do not capture large volumes of CO_2 for CCS.

Several projects in the United States and abroad—typically associated with oil and gas production—are successfully capturing, injecting, and storing CO_2 underground, albeit at relatively small scales. The oil and gas industry in the United States injects nearly 50 million tons of CO_2 underground each year for the purpose of enhanced oil recovery (EOR). The volume of CO_2 envisioned for CCS as a climate

---

[*] This is an edited, reformatted and augmented version of a Congressional Research Service publication, CRS Report for Congress R42532, from www.crs.gov, prepared for Members and Committees of Congress, dated May 14, 2012.

mitigation option is overwhelming compared to the amount of CO_2 used for EOR.

According to the U.S. Department of Energy (DOE), the United States has the potential to store billions of tons of CO_2 underground and keep the gas trapped there indefinitely. Capturing and storing the equivalent of decades or even centuries of CO_2 emissions from power plants (at current levels of emissions) suggests that CCS has the potential to reduce U.S. greenhouse gas emissions substantially while allowing the continued use of fossil fuels.

An integrated CCS system would include three main steps: (1) capturing and separating CO_2 from other gases; (2) purifying, compressing, and transporting the captured CO_2 to the sequestration site; and (3) injecting the CO_2 in subsurface geological reservoirs or storing it in the oceans. Deploying CCS technology on a commercial scale would be a vast undertaking.

The CCS process, although simple in concept, would require significant investments of capital and of time.

Capital investment would be required for the technology to capture CO_2 and for the pipeline network to transport the captured CO_2 to the disposal site.

Time would be required to assess the potential CO_2 storage reservoir, inject the captured CO_2, and monitor the injected plume to ensure against leaks to the atmosphere or to underground sources of drinking water, potentially for years or decades until injection activities cease and the injected plume stabilizes.

Three main types of geological formations in the United States are being considered for storing large amounts of CO_2: oil and gas reservoirs, deep saline reservoirs, and unmineable coal seams. The deep ocean also has a huge potential to store carbon; however, direct injection of CO_2 into the deep ocean is controversial, and environmental concerns have forestalled planned experiments in the open ocean. Mineral carbonation—reacting minerals with a stream of concentrated CO_2 to form a solid carbonate—is well understood, but it is still an experimental process for storing large quantities of CO_2.

Large-scale CCS injection experiments are only beginning in the United States to test how different types of reservoirs perform during CO_2 injection of a million tons of CO_2 or more. Results from the experiments will undoubtedly be crucial to future permitting and site approval regulations.

Acceptance by the general public of large-scale deployment of CCS may be a significant challenge. Some of the large-scale injection tests could garner information about public acceptance, as citizens become familiar with the concept, process, and results of CO_2 injection tests in their local communities.

INTRODUCTION

Carbon capture and sequestration (or storage)—known as CCS—is a physical process that involves capturing manmade carbon dioxide (CO_2) at its source and storing it before its release to the atmosphere. CCS could reduce the amount of CO_2 emitted to the atmosphere despite the continued use of fossil fuels. An integrated CCS system would include three main steps:
(1) capturing CO_2 and separating it from other gases; (2) purifying, compressing, and transporting the captured CO_2 to the sequestration site; and (3) injecting the CO_2 in subsurface geological reservoirs or storing it in the oceans. As a measure for mitigating global climate change, CCS has attracted congressional interest and support because several projects in the United States and abroad—typically associated with oil and gas production—are successfully capturing, injecting, and storing CO_2 underground, albeit at relatively small scales. The oil and gas industry in the United States injects approximately 48 million metric tons of CO_2 underground each year to help recover oil and gas resources (a process known as enhanced oil recovery, or EOR).[1] Potentially, much larger amounts of CO_2 produced from electricity generation—approximately 2.2 billion metric tons per year, over 40% of the total CO_2 emitted in the United States from fossil fuels (see *Table 1*)—could be targeted for large-scale CCS.

Table 1. Sources for CO_2 Emissions in the United States from Combustion of Fossil Fuels

Sources	CO_2 Emissions[a] (millions of metric tons)	Percent of Total
Electricity generation	2,216.8	42%
Transportation	1,745.5	33%
Industrial	777.8	15%
Residential	340.2	6%
Commercial	224.2	4%
Total	5,304.5	100%

Source: U.S. Environmental Protection Agency (EPA), Inventory of U.S. Greenhouse Emissions and Sinks: 1990- 2010, Table ES-3 (2012); see http://epa.gov/climate change/emissions/usinventoryreport.html.

[a] CO_2 emissions in millions of metric tons for 2010; excludes emissions from U.S. territories.

Fuel combustion accounts for 94.4% of all U.S. CO_2 emissions.[2] Electricity generation contributes the largest proportion of CO_2 emissions compared to other types of fossil fuel use in the United States (*Table 1*). Electricity-generating plants are among the most likely initial candidates for capture, separation, and storage or reuse of CO_2 because they are predominantly large, stationary, single-point sources of emissions. Large industrial facilities, such as cement-manufacturing, ethanol, or hydrogen production plants, that produce large quantities of CO_2 as part of the industrial process are also good candidates for CO_2 capture and storage.[3]

This report is a brief summary of what CCS is, how it is supposed to work, why it has gained the interest and support of some members of Congress, and what some of the challenges are to its implementation and deployment across the United States.

This report covers only CCS and not other types of carbon sequestration activities whereby CO_2 is removed from the atmosphere and stored in vegetation or soils, such as forests and agricultural lands.[4]

CO_2 CAPTURE

The first step in CCS is to capture CO_2 at the source and produce a concentrated stream for transport and storage. Currently, three main approaches are available to capture CO_2 from large-scale industrial facilities or power plants: (1) post-combustion capture, (2) pre-combustion capture, and (3) oxy-fuel combustion capture. For power plants, current commercial CO_2 capture systems could operate at 85%-95% capture efficiency.[5] The capture phase of the CCS process, however, may be 80% or more of the total costs for CCS.[6]

Post-Combustion Capture

This process involves extracting CO_2 from the flue gas following combustion of fossil fuels or biomass. Several commercially available technologies, some involving absorption using chemical solvents, can in principle be used to capture large quantities of CO_2 from flue gases. U.S. commercial electricity-generating plants currently do not capture large volumes of CO_2 because they are not required to and there are no economic incentives to do so. Nevertheless, the post-combustion capture process includes proven technologies that are commercially available today.

Pre-Combustion Capture

This process separates CO_2 from the fuel by combining the fuel with air and/or steam to produce hydrogen for combustion and a separate CO_2 stream that could be stored. The most common technologies today use steam reforming, in which steam is employed to extract hydrogen from natural gas.[7]

Oxy-Fuel Combustion Capture

This process uses oxygen instead of air for combustion and produces a flue gas that is mostly CO_2 and water, which are easily separable, after which the CO_2 can be compressed, transported, and stored. The U.S. Department of Energy's (DOE) flagship CCS demonstration project, FutureGen, plans to retrofit an existing power unit with an oxy-fuel combustion unit.[8]

CO_2 TRANSPORT

Pipelines are the most common method for transporting CO_2 in the United States. Currently, approximately 4,100 miles of pipeline transport CO_2 in the United States, predominately to oil and gas fields, where it is used for EOR.[9] Transporting CO_2 in pipelines is similar to transporting petroleum products like natural gas and oil; it requires attention to design, monitoring for leaks, and protection against overpressure, especially in populated areas.[10]

Using ships may be feasible when CO_2 must be transported over large distances or overseas. Ships transport CO_2 today, but at a small scale because of limited demand. Liquefied natural gas, propane, and butane are routinely shipped by marine tankers on a large scale worldwide. Rail cars and trucks can also transport CO_2, but this mode would probably be uneconomical for large-scale CCS operations.

Costs for pipeline transport vary, depending on construction, operation and maintenance, and other factors, including right-of-way costs, regulatory fees, and more. The quantity and distance transported will mostly determine costs, which will also depend on whether the pipeline is onshore or offshore, the level of congestion along the route, and whether mountains, large rivers, or frozen ground are encountered. Shipping costs are unknown in any detail, however, because no large-scale CO_2 transport system (in millions of metric tons of CO_2 per year, for example) is operating. Ship costs might be lower than pipeline transport for distances greater than 1,000 kilometers and for less than a few million metric tons of CO_2 ($MtCO_2$)[11] transported per year.[12]

Even though regional CO_2 pipeline networks currently operate in the United States for enhanced EOR, developing a more expansive network for CCS could pose numerous regulatory and economic challenges. Some of these include questions about pipeline network requirements, economic regulation, utility cost recovery, regulatory classification of CO_2 itself, and pipeline safety.[13]

CO$_2$ SEQUESTRATION

Three main types of geological formations are being considered for carbon sequestration: (1) depleted oil and gas reservoirs, (2) deep saline reservoirs, and (3) unmineable coal seams. In each case, CO_2 would be injected in a supercritical state—a relatively dense liquid—below ground into a porous rock formation that holds or previously held fluids. When CO_2 is injected at depths greater than 800 meters in a typical reservoir, the pressure keeps the injected CO_2 in a supercritical state (dense like a liquid, fluid like a gas) and thus it is less likely to migrate out of the geological formation. Injecting CO_2 into deep geological formations uses existing technologies that have been primarily developed and used by the oil and gas industry, and that could potentially be adapted for long-term storage and monitoring of CO_2. Other underground injection applications in practice today, such as natural gas storage, deep injection of liquid wastes, and subsurface disposal of oil-field brines, can also provide valuable experience and information for sequestering CO_2 in geological formations.[14]

The storage capacity for CO_2 storage in geological formations is potentially huge if all the sedimentary basins in the world are considered (see discussion below of storage capacity estimates for the United States).[15] The suitability of any particular site, however, depends on many factors, including proximity to CO_2 sources and other reservoir-specific qualities like porosity, permeability, and potential for leakage. For CCS to succeed, it is assumed that each reservoir type would permanently store the vast majority of injected CO_2, keeping the gas isolated from the atmosphere in perpetuity.

Oil and Gas Reservoirs

Pumping CO_2 into oil and gas reservoirs to boost production (EOR) is practiced in the petroleum industry today. The United States is a world leader in this technology, and oil and gas operators inject approximately 48 $MtCO_2$ underground each year to help recover oil and gas resources.[16] Most of the CO_2 used for EOR in the United States comes from naturally occurring geologic formations, however, not from industrial sources. Using CO_2 from industrial emitters has appeal because the costs of capture and transport from the facility could be partially offset by revenues from oil and gas production.

Carbon dioxide can be used for EOR onshore or offshore. To date, most CO_2 projects associated with EOR are onshore, with the bulk of U.S. activities in west Texas. Carbon dioxide can also be injected into oil and gas reservoirs that are completely depleted, which would serve the purpose of long-term sequestration, but without any offsetting benefit from oil and gas production.

Advantages and Disadvantages

Depleted or abandoned oil and gas fields, especially in the United States, are considered prime candidates for CO_2 storage for several reasons:

- oil and gas originally trapped did not escape for millions of years, demonstrating the structural integrity of the reservoir;
- extensive studies for oil and gas typically have characterized the geology of the reservoir;
- computer models have often been developed to understand how hydrocarbons move in the reservoir, and the models could be applied to predicting how CO_2 could move; and
- infrastructure and wells from oil and gas extraction may be in place and might be used for handling CO_2 storage.

Some of these features could also be disadvantages to CO_2 sequestration. Wells that penetrate from the surface to the reservoir could be conduits for CO_2 release if they are not plugged properly. Care must be taken not to overpressure the reservoir during CO_2 injection, which could fracture the caprock—the part of the formation that formed a seal to trap oil and gas—and subsequently allow CO_2 to escape. Also, shallow oil and gas fields (those less than 800 meters deep, for example) may be unsuitable because CO_2 may form a gas instead of a denser liquid and could escape to the surface more easily. In addition, oil and gas fields that are suitable for EOR may not necessarily be located near industrial sources of CO_2. Costs to construct pipelines to connect sources of CO_2 with oil and gas fields may, in part, determine whether an EOR operation using industrial sources of CO_2 is feasible.

Although the United States injects nearly 50 $MtCO_2$ underground each year for the purposes of EOR, that amount represents approximately 2% of the CO_2 emitted from fossil fuel electricity generation alone. The sheer volume of CO_2 envisioned for CCS as a climate mitigation option is overwhelming compared to the amount of CO_2 used for EOR. It may be that EOR will increase in the future, depending on economic, regulatory, and technical factors, and more CO_2 will be sequestered as a consequence. It is also likely that EOR would only account for a small fraction of the total amount of CO_2 injected underground in the future, even if CCS becomes a significant component in an overall scheme to substantially reduce CO_2 emissions to the atmosphere.

The In Salah and Weyburn Projects

The In Salah Project in Algeria is the world's first large-scale effort to store CO_2 in a natural gas reservoir.[17] At In Salah, CO_2 is separated from the produced natural gas (the gas contains approximately 5.5% CO_2) and then reinjected into the same formation. Approximately 17 $MtCO_2$ are planned to be captured and stored over the lifetime of the project at a rate of slightly more than 1 Mt per year.[18]

The Weyburn Project in south-central Canada uses CO_2 produced from a coal gasification plant in North Dakota for EOR, injecting up to 5,000 tCO_2 per day into the formation and recovering oil.[19] Approximately 20 $MtCO_2$ are expected to remain in the formation over the lifetime of the project.[20]

Deep Saline Reservoirs

Some rocks in sedimentary basins contain saline fluids—brines or brackish water unsuitable for agriculture or drinking. As with oil and gas, deep saline reservoirs can be found onshore and offshore; in fact, they are often part of oil and gas reservoirs and share many characteristics. The oil industry routinely injects brines recovered during oil production into saline reservoirs for disposal.[21] Using suitably deep saline reservoirs for CO_2 sequestration has advantages: (1) they are more widespread in the United States than oil and gas reservoirs and thus have greater probability of being close to large point sources of CO_2; and (2) saline reservoirs have potentially the largest reservoir capacity of the three types of geologic formations.

Advantages and Disadvantages

Although deep saline reservoirs potentially have huge capacity to store CO_2, estimates of lower and upper capacities vary greatly, reflecting a higher degree of uncertainty in how to measure storage capacity.[22] Actual storage capacity may have to be determined on a case-by-case basis. Estimates of storage capacity for the United States from the DOE Regional Sequestration Partnership Program are discussed below.

From estimates of the potential storage capacity in saline reservoirs, it is likely that the vast majority of CO_2 injected underground would be stored in these formations, assuming that CCS were deployed on a commercial scale across the United States. In addition to their potential capacity, deep saline reservoirs underlie large portions of the country, and could be more easily accessible to large, stationary sources of CO_2 than oil and gas reservoirs or coal seams. *Figure 1* shows broad outlines of sedimentary basins containing the deep saline reservoirs, and the locations of a variety of stationary sources of CO_2.

The DOE Regional Sequestration Partnership Program has conducted simulations, field studies, small-scale injection projects, and is now beginning a phase of large-scale injection demonstration projects to investigate the suitability of deep saline reservoirs.[23] Because of the potentially vast amounts of CO_2 that could be sequestered, these experiments could shed light on the potential for leakage of CO_2 from the reservoir, and test the ability to detect the movement of CO_2 underground as well as to detect leaks through overlying cap rocks.

In addition to the possibility of CO_2 leakage, injection of millions of tons of CO_2 will displace large volumes of brine in the deep saline reservoirs. One disadvantage is therefore the possibility that displaced brine could leak into underground sources of drinking water. Ultimately, CO_2 will likely dissolve into the brine, but that could take decades.

Also, injecting large volumes of fluid into the subsurface has the potential to trigger earthquakes, especially if the CO_2 is injected into an undetected fault. Presumably, evaluating the potential storage site prior to beginning injection will limit the potential for triggering earthquakes (also referred to as "induced seismicity"), but there is no guarantee that fluid could not migrate to faulted or fractured rocks over the course of many years and induce an earthquake.

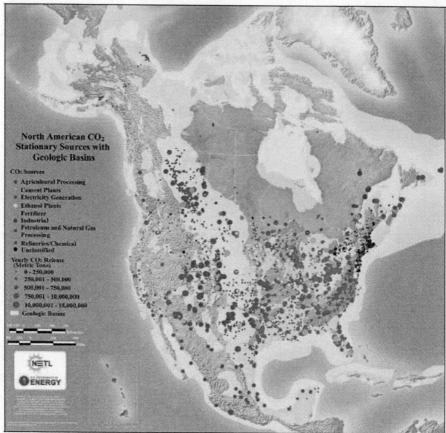

Source: U.S. Department of Energy, National Energy Technology Laboratory, "2010
   Carbon Sequestration Atlas of the United States and Canada, Third Edition,"
   http://www.netl.doe.gov/technologies/carbon_seq/refshelf/ atlasIII/index.html.
Note: Not all geologic basins have deep saline reservoirs suitable for carbon
   sequestration.

Figure 1. Stationary Sources of CO2 in North America and Underlying Geologic
Basins.

The issue of induced seismicity has recently been linked to the injection
and disposal of produced waters from oil and gas fields.[24]

Unlike sequestration in existing oil and gas fields, injecting into deep
saline reservoirs may take place in regions of the country that have not
experienced drilling activities. Public opposition may arise to activities on the
surface—such as land clearing, building of new roads, transport of heavy
equipment, and operation of drill rigs—but also to the concept of disposing

CO_2 underground near residences and communities. There is at least one example of public opposition to CO_2 injection leading to cancellation of a project in Europe.[25]

The Sleipner Project

The Sleipner Project in the North Sea is the first commercial-scale operation for sequestering CO_2 in a deep saline reservoir. The Sleipner project has been operating since 1996, and it injects and stores approximately 2,800 tCO_2 per day, or about 1 $MtCO_2$ per year.[26] Carbon dioxide is separated from natural gas production at the nearby Sleipner West Gas Field, compressed, and then injected 800 meters below the seabed of the North Sea into the Utsira formation, a sandstone reservoir 200-250 meters (650-820 feet) thick containing saline fluids. Monitoring has indicated the CO_2 has not leaked from the saline reservoir, and computer simulations suggest that the CO_2 will eventually dissolve into the saline water, reducing the potential for leakage in the future.

Another CO_2 sequestration project, similar to Sleipner, began in the Barents Sea in April 2008 (the Snohvit Project),[27] and is injecting approximately 2,000 tCO_2 per day below the seafloor. A larger project is being planned in western Australia (the Gorgon Project)[28] that would inject 9,000 tCO_2 per day when at full capacity. Similar to the Sleipner and Snohvit operations, the Gorgon plans to strip CO_2 from produced natural gas and inject it into deep saline formations for permanent storage.

Unmineable Coal Seams

U.S. coal resources not mineable with current technology are those where the coal beds[29] are not thick enough, or are too deep, or whose structural integrity is inadequate for mining. Even if they cannot be mined, coal beds are commonly permeable and can trap gases, such as methane, which can be extracted (a resource known as coal-bed methane, or CBM). Methane and other gases are physically bound (adsorbed) to the coal. Studies indicate that CO_2 binds even more tightly to coal than methane.[30] Carbon dioxide injected into permeable coal seams could displace methane, which could be recovered by wells and brought to the surface, providing a source of revenue to offset the costs of CO_2 injection.

Advantages and Disadvantages

Unmineable coal seam injection projects would need to assess several factors in addition to the potential for CBM extraction. These include depth, permeability, coal bed geometry (a few thick seams, not several thin seams), lateral continuity and vertical isolation (less potential for upward leakage), and other considerations. Once CO_2 is injected into a coal seam, it would likely remain there unless the seam is depressurized or the coal is mined. Many unmineable coal seams in the United States are located relatively near electricity-generating facilities, which could reduce the distance and cost of transporting CO_2 from large point sources to storage sites.

Not all types of coal beds are suitable for CBM extraction. Without the coal-bed methane resource, the sequestration process would be less economically attractive. However, the displaced methane would need to be combusted or captured because methane itself is a more potent greenhouse gas than CO_2. Once burned, methane produces mostly CO_2 and water.

Without ongoing commercial experience, storing CO_2 in coal seams has significant uncertainties compared to the other two types of geological storage discussed. According to IPCC, unmineable coal seams have the smallest potential capacity for storing CO_2 globally compared to oil and gas fields or deep saline formations. The latest assessment from DOE also indicates that unmineable coal seams in the United States have less potential capacity than U.S. oil and gas fields for storing CO_2. (See following discussion.) No commercial CO_2 injection and sequestration projects in coal beds are currently underway in the United States.

GEOLOGICAL STORAGE CAPACITY FOR CO_2 IN THE UNITED STATES

As *Figure 1* indicates, geologic basins containing at least one of each of these three types of potential CO_2 reservoirs occur across most of the United States, in relative proximity to many large point sources of CO_2, such as fossil fuel power plants or cement plants. The DOE Regional Sequestration Partnership Program has produced estimates of the potential storage capacity for each of these types of reservoirs and published the estimates in a Carbon Sequestration Atlas. The 2010 Carbon Sequestration Atlas (third edition) updates the 2008 version (second edition), and a summary of the storage estimates for both editions is compared in *Table 2*.[31]

The Carbon Sequestration Atlas was compiled from estimates of geological storage capacity made by seven separate regional partnerships (government-industry collaborations fostered by DOE) that each produced estimates for different regions of the United States and parts of Canada.

According to DOE, geographical differences in fossil fuel use and sequestration potential across the country led to a regional approach to assessing CO_2 sequestration potential.[32] The Carbon Sequestration Atlas reflects some of the regional differences; for example, not all of the regional partnerships identified unmineable coal seams as potential CO_2 reservoirs. Other partnerships identified geological formations unique to their regions— such as organic-rich shales in the Illinois Basin, or flood basalts in the Columbia River Plateau—as other types of possible reservoirs for CO_2 storage.

Table 2 indicates a lower and upper range for sequestration potential in deep saline formations and for unmineable coal seams, but only a single estimate for oil and gas fields. Comparison between the 2008 and 2010 estimates indicates small changes between the two estimates for oil and gas fields, but relatively larger changes in estimates for deep saline formations and unmineable coal seams. It is clear from the table that DOE considers estimates for oil and gas fields much better constrained than for the other types of reservoirs. The amount and types of data from oil and gas fields, such as production history, and reservoir volume calculations, often represent decades of experience in the oil and gas industry. In the Carbon Sequestration Atlas, oil and gas reservoirs were assessed at the field level (i.e., on a finer scale and in more detail) than deep saline formations or unmineable coal seams, which were assessed at the basin level (i.e., at a coarser scale and in less detail).

Other methodologies for and estimates of the geological sequestration potential have been released or are underway. For example, the Energy Independence and Security Act of 2007 (EISA, P.L. 110-140) directed the Department of the Interior (DOI) to develop a single methodology for an assessment of the national potential for geologic storage of carbon dioxide. The U.S. Geological Survey (USGS) released an initial methodology in 2009. In response to external comments and reviews, the USGS revised its initial methodology in a 2010 report. [33] According to DOE, the USGS effort will allow refinement of the estimates provided in the 2008 Carbon Sequestration Atlas, and will incorporate uncertainty in the capacity estimates.[34] In addition, DOE notes that its methodology will incorporate results from large-scale carbon sequestration demonstration projects now underway, and that it will update its CO_2 storage estimates every two years. The DOE Sequestration

Atlas should probably be considered an evolving assessment of U.S. reservoir capacity for CO_2 storage.[35]

The total lower estimate (sum of the three reservoir types) from the 2010 Carbon Sequestration Atlas shown in *Table 2* indicates the potential to store the equivalent of 830 years of CO_2 emissions from electricity generation in the United States at current emission rates (2.2 billion tons per year). The total upper estimate indicates the potential for over 9,000 years of CO_2 emissions from electricity generation.

Table 2. Geological Sequestration Potential for the United States and Parts of Canada (billion metric tons of CO_2)

Reservoir type	Lower estimate (2010)	Lower estimate (2008)	% Change	Upper estimate (2010)	Upper estimate (2008)	% Change
Oil and gas fields	143	138	+3.6%	143	138	+3.6%
Deep saline formations	1,653	3,297	-50%	20,213	12,618	+60%
Unmineable coal seams	60	157	-62%	117	178	-34%
Totals	1,856	3,592	-48%	20,473	12,934	+58%

Source: 2008 and 2010 Carbon Sequestration Atlases.

DEEP OCEAN SEQUESTRATION

The world's oceans contain approximately 50 times the amount of carbon stored in the atmosphere and nearly 10 times the amount stored in plants and soils.[36] The oceans today take up—act as a net sink for—approximately 1.7 $GtCO_2$ per year. About 45% of the CO_2 released from fossil fuel combustion and land use activities during the 1990s has remained in the atmosphere, while the remainder has been taken up by the oceans, vegetation, or soils on the land surface.[37] Without the ocean sink, atmospheric CO_2 concentration would be increasing more rapidly. Ultimately, the oceans could store more than 90% of all the carbon released to the atmosphere by human activities, but the process takes thousands of years.[38] The ocean's capacity to absorb atmospheric CO_2 may change, however, and possibly even decrease in the future.[39] Also, studies indicate that as more CO_2 enters the ocean from the atmosphere, the surface waters are becoming more acidic.[40]

Advantages and Disadvantages

Although the surface of the ocean is becoming more concentrated with CO_2, the surface waters and the deep ocean waters generally mix very slowly, on the order of decades to centuries. Injecting CO_2 directly into the deep ocean would take advantage of the slow rate of mixing, allowing the injected CO_2 to remain sequestered until the surface and deep waters mix and CO_2 concentrations equilibrate with the atmosphere. What happens to the CO_2 would depend on how it is released into the ocean, the depth of injection, and the temperature of the seawater.

Carbon dioxide injected at depths shallower than 500 meters typically would be released as a gas, and would rise towards the surface. Most of it would dissolve into seawater if the injected CO_2 gas bubbles were small enough.[41] At depths below 500 meters, CO_2 can exist as a liquid in the ocean, although it is less dense than seawater. After injection below 500 meters, CO_2 would also rise, but an estimated 90% would dissolve in the first 200 meters. Below 3,000 meters in depth, CO_2 is a liquid and is denser than seawater; the injected CO_2 would sink and dissolve in the water column or possibly form a CO_2 pool or lake on the sea bottom. Some researchers have proposed injecting CO_2 into the ocean bottom sediments below depths of 3,000 meters, and immobilizing the CO_2 as a dense liquid or solid CO_2 hydrate.[42] Deep storage in ocean bottom sediments, below 3,000 meters in depth, might potentially sequester CO_2 for thousands of years.[43]

The potential for ocean storage of captured CO_2 is huge, but environmental impacts on marine ecosystems and other issues may determine whether large quantities of captured CO_2 will ultimately be stored in the oceans. Also, deep ocean storage is in a research stage, and the effects of scaling up from small research experiments, using less than 100 liters of CO_2,[44] to injecting several $GtCO_2$ into the deep ocean are unknown.

Injecting CO_2 into the deep ocean would change ocean chemistry, locally at first, and assuming that hundreds of $GtCO_2$ were injected, would eventually produce measurable changes over the entire ocean.[45] The most significant and immediate effect would be the lowering of pH, increasing the acidity of the water. A lower pH may harm some ocean organisms, depending on the magnitude of the pH change and the type of organism. Actual impacts of deep sea CO_2 sequestration are largely unknown, however, because scientists know very little about deep ocean ecosystems.[46]

Environmental concerns led to the cancellation of the largest planned experiment to test the feasibility of ocean sequestration in 2002. A scientific consortium had planned to inject 60 tCO_2 into water over 800 meters deep near

the Kona coast on the island of Hawaii. Environmental organizations opposed the experiment on the grounds that it would acidify Hawaii's fishing grounds, and that it would divert attention from reducing greenhouse gas emissions.[47] A similar but smaller project with plans to release more than 5 tCO_2 into the deep ocean off the coast of Norway, also in 2002, was cancelled by the Norway Ministry of the Environment after opposition from environmental groups.[48]

Sequestering Under the Seabed

Deep ocean sequestration, as discussed here, is different from injecting CO_2 beneath the seabed into depleted oil and gas reservoirs or deep saline formations.

The Sleipner project discussed above is an example of injection beneath the seafloor, but not injection into the ocean waters. Sequestering CO_2 under the seabed on the U.S. continental shelf would eliminate the need to negotiate with local landowners over the rights to surface land and to the pore space in the subsurface.

However, it would also require developing an offshore infrastructure to transport and inject the captured CO_2, along with all the other challenges of evaluating the potential offshore reservoir, including monitoring the injected CO_2, and providing for liability and ownership of the CO_2 after injection has ceased.

MINERAL CARBONATION

Another option for sequestering CO_2 produced by fossil fuel combustion involves converting CO_2 to solid inorganic carbonates, such as $CaCO_3$ (limestone), using chemical reactions. When this process occurs naturally, it is known as "weathering" and takes place over thousands or millions of years. The process can be accelerated by reacting a high concentration of CO_2 with minerals found in large quantities on the Earth's surface, such as olivine or serpentine.[49] Mineral carbonation has the advantage of sequestering carbon in solid, stable minerals that can be stored without risk of releasing carbon to the atmosphere over geologic time scales.[50]

Mineral carbonation involves three major activities: (1) preparing the reactant minerals—mining, crushing, and milling—and transporting them to a processing plant, (2) reacting the concentrated CO_2 stream with the prepared minerals, and (3) separating the carbonate products and storing them in a suitable repository.

Advantages and Disadvantages

Mineral carbonation is well understood and can be applied at small scales, but is at an early phase of development as a technique for sequestering large amounts of captured CO_2. Large volumes of silicate oxide minerals are needed, from 1.6 to 3.7 metric tons of silicates per tCO_2 sequestered. Thus, a large-scale mineral carbonation process needs a large mining operation to provide the reactant minerals in sufficient quantity.[51] Large volumes of solid material would also be produced, between 2.6 and 4.7 metric tons of materials per tCO_2 sequestered, or 50%-100% more material to be disposed of by volume than originally mined. Because mineral carbonation is in the research and experimental stage, estimating the amount of CO_2 that could be sequestered by this technique is difficult.

One possible type of geological reservoir for CO_2 storage—major flood basalts[52] such as those on the Columbia River Plateau—is being explored for its potential to react with CO_2 and form solid carbonates in situ (in place). Instead of mining, crushing, and milling the reactant minerals, as discussed above, CO_2 would be injected directly into the basalt formations and would react with the rock over time and at depth to form solid carbonate minerals. Large and thick formations of flood basalts occur globally, and many have characteristics—such as high porosity and permeability—that are favorable to storing CO_2. Those characteristics, combined with the tendency of basalt to react with CO_2, could result in a large-scale conversion of the gas into stable, solid minerals that would remain underground for geologic time. The DOE regional carbon sequestration partnerships are exploring the possibility of using Columbia River Plateau flood basalts in the Pacific Northwest for storing CO_2.[53]

CURRENT ISSUES AND FUTURE CHALLENGES

A primary goal of developing and deploying CCS is to allow large industrial facilities, such as fossil fuel power plants and cement plants, to operate while reducing their CO_2 emissions by 80%-90%. Such reductions would presumably reduce the likelihood of continued climate warming from greenhouse gases by slowing the rise in atmospheric concentrations of CO_2. To achieve the overarching goal of reducing the likelihood of continued climate warming would depend, in part, on how fast and how widely CCS could be deployed throughout the economy.

The additional cost of installing CCS on CO_2-emitting facilities is a primary challenge to the adoption and deployment of CCS in the United States.

Major increases in CO_2 capture technology efficiency will likely produce the greatest relative cost savings for CCS systems, but challenges also face the transport and storage components of CCS. Ideally, storage reservoirs for CO_2 would be located close to sources, obviating the need to build a large pipeline infrastructure to deliver captured CO_2 for underground sequestration. If CCS moves to widespread implementation, however, some areas of the country may not have adequate reservoir capacity nearby, and may need to construct pipelines from sources to reservoirs. Identifying and validating sequestration sites would need to account for CO_2 pipeline costs, for example, if the economics of the sites are to be fully understood. If this is the case, there would be questions to be resolved regarding pipeline network requirements, economic regulation, utility cost recovery, regulatory classification of CO_2 itself, and pipeline safety. In addition, Congress may be called upon to address federal jurisdictional authority over CO_2 pipelines under existing law, and whether additional legislation may be necessary if a CO_2 pipeline network grows and crosses state lines.

Although DOE has identified substantial potential storage capacity for CO_2, particularly in deep saline formations, large-scale injection experiments are only beginning in the United States to test how different types of reservoirs perform during CO_2 injection. Data from the experiments will undoubtedly be crucial to future permitting and site approval regulations.

In addition, liability, ownership, and long-term stewardship for CO_2 sequestered underground are issues that would need to be resolved before CCS is deployed commercially.

Some states are moving ahead with state-level geological sequestration regulations for CO_2, so federal efforts to resolve these issues at a national level would likely involve negotiations with the states. Acceptance by the general public of large-scale deployment of CCS may be a significant challenge if the majority of CCS projects involve private land. Some of the large-scale injection tests could garner information about public acceptance, as local communities become familiar with the concept, process, and results of CO_2 injection tests. Apart from the question of how the public would accept the likely higher cost for electricity generated from plants with CCS, how a growing CCS infrastructure of pipelines, injection wells, underground reservoirs, and other facilities would be accepted by the public is as yet unknown.

End Notes

[1] U.S. Department of Energy, National Energy Technology Laboratory, Carbon Sequestration Through Enhanced Oil Recovery, (March, 2008), at http://www.netl.doe.gov/publications/factsheets/program/Prog053.pdf.

[2] U.S. Environmental Protection Agency (EPA), Inventory of U.S. Greenhouse Emissions and Sinks: 1990-2010, p. ES-7. The percentage refers to U.S. emissions in 2010; see http://epa.gov/climatechange/emissions/usinventoryreport.html.

[3] Intergovernmental Panel on Climate Change (IPCC) Special Report: Carbon Dioxide Capture and Storage, 2005. (Hereafter referred to as IPCC Special Report.)

[4] For more information about carbon sequestration in forests and agricultural lands, see CRS Report RL31432, Carbon Sequestration in Forests, by Ross W. Gorte; CRS Report RL33898, Climate Change: The Role of the U.S. Agriculture Sector, by Renée Johnson; and CRS Report R40186, Biochar: Examination of an Emerging Concept to Sequester Carbon, by Kelsi Bracmort. For more information about carbon exchanges between the oceans, atmosphere, and land surface, see CRS Report RL34059, The Carbon Cycle: Implications for Climate Change and Congress, by Peter Folger.

[5] IPCC Special Report, p. 107.

[6] See, for example, John Deutch et al., The Future of Coal, Massachusetts Institute of Technology, An Interdisciplinary MIT Study, 2007, Executive Summary, p. xi.

[7] See CRS Report R41325, Carbon Capture: A Technology Assessment, by Peter Folger.

[8] See CRS Report R42496, Carbon Capture and Sequestration: Research, Development, and Demonstration at the U.S. Department of Energy, by Peter Folger, for further discussion of FutureGen.

[9] Kevin Bliss et al., "A Policy, Legal, and Regulatory Evaluation of the Feasibility of a National Pipeline Infrastructure for the Transport and Storage of Carbon Dioxide," Interstate Oil and Gas Compact Commission, September 10, 2010, Table 3, http://www.sseb.org/ downloads/pipeline.pdf. By comparison, nearly 500,000 miles of pipeline operate to convey natural gas and hazardous liquids in the United States.

[10] IPCC Special Report, p. 181.

[11] One metric ton of CO_2 equivalent is written as 1 tCO_2; one million metric tons is written as 1 MtCO2; one billion metric tons is written as 1 $GtCO_2$.

[12] IPCC Special Report, p. 31.

[13] These issues are discussed in more detail in CRS Report RL33971, Carbon Dioxide (CO_2) Pipelines for Carbon Sequestration: Emerging Policy Issues, by Paul W. Parfomak, Peter Folger, and Adam Vann, and CRS Report RL34316, Pipelines for Carbon Dioxide (CO_2) Control: Network Needs and Cost Uncertainties, by Paul W. Parfomak and Peter Folger.

[14] IPCC Special Report, p. 31.

[15] Sedimentary basins refer to natural large-scale depressions in the Earth's surface that are filled with sediments and fluids and are therefore potential reservoirs for CO_2 storage.

[16] Data from 2006. See DOE, National Energy Technology Laboratory, Carbon Sequestration Through Enhanced Oil Recovery, (March 2008), at http://www.netl.doe.gov/publications/factsheets/program/Prog053.pdf.

[17] IPCC Special Report, p. 203.

[18] The Carbon Capture and Sequestration Technologies Program at MIT, Carbon Capture and Sequestration Project Database, In Salah Fact Sheet, http://sequestration.mit.edu/tools/projects/in_salah.html.

[19] IPCC Special Report, p. 204.

[20] MIT Carbon Capture and Sequestration Project Database, Weyburn Fact Sheet, http://sequest ration.mit.edu/tools/ projects/weyburn.html.

[21] DOE Office of Fossil Energy; see http://www.fossil.energy.gov/programs/ sequestration/ geolo gic/index.html.

[22] IPCC Special Report, p. 223.

[23] See CRS Report R42496, Carbon Capture and Sequestration: Research, Development, and Demonstration at the U.S. Department of Energy, by Peter Folger, for more information on the DOE programs.

[24] See, for example, Mike Soraghan, "Drilling Waste Disposal Risks Another Damaging Okla. Quake, Scientist Warns," Energywire, April 19, 2012, http://www.eenews.net/energywire/ 2012/04/19/archive/1?terms=earthquake.

[25] See Paul Voosen, "Public Outcry Scuttles German Demonstration Plant," Greenwire, December 6, 2011, http://www.eenews.net/Greenwire/2011/12/06/ archive/10?terms= vatt enfall.

[26] Carbon Capture and Sequestration Project Database, Sleipner Fact Sheet, http://sequestration. mit.edu/tools/projects/ sleipner.html.

[27] Carbon Capture and Sequestration Project Database, Snohvit Fact Sheet, http://sequest ration.mit.edu/tools/projects/ snohvit.html

[28] Carbon Capture and Sequestration Project Database, Gorgon Fact Sheet, http://sequestration. mit.edu/tools/projects/ gorgon.html.

[29] Coal bed and coal seam are interchangeable terms.

[30] IPCC Special Report, p. 217.

[31] U.S. Dept. of Energy, National Energy Technology Laboratory, 2010 Carbon Sequestration Atlas of the United States and Canada, 3rd ed. (November 2010), 160 pages. Hereinafter referred to as the 2010 Carbon Sequestration Atlas, http://www.netl.doe.gov/technologies/ carbon_seq/refshelf/atlasIII/2010atlasIII.pdf; and 2008 Carbon Sequestration Atlas of the United States and Canada, 2nd ed. (November 2008), 140 pages. Hereinafter referred to as the 2008 Carbon Sequestration Atlas. (Available from CRS.) A 2007 Carbon Sequestration Atlas was also published and is available from CRS.

[32] 2008 Carbon Sequestration Atlas, p. 8.

[33] Sean T. Brennan et al., A Probabilistic Assessment Methodology for the Evaluation of Geologic Carbon Dioxide Storage, USGS, Open-File Report 2010-1127, 2010. The USGS has also released at least one report on the geologic framework of specific geologic basins that may help improve estimates of sequestration capacity. See, for example, Jacob A. Covault et al., Geologic Framework for the National Assessment of Carbon Dioxide Storage Resources-Bighorn Basin, Wyoming and Montana, USGS, Open-File Report 2012-1024-A, 2012, http://pubs.usgs.gov/of/2012/ 1024/a/.

[34] 2008 Carbon Sequestration Atlas, p. 23.

[35] 2010 Carbon Sequestration Atlas, p. 139.

[36] Christopher L. Sabine et al., "Current Status and Past Trends of the Global Carbon Cycle," in C. B. Field and M. R. Raupach, eds., The Global Carbon Cycle: Integrating Humans, Climate, and the Natural World (Washington, DC: Island Press, 2004), pp. 17-44.

[37] 2007 IPCC Working Group I Report, pp. 514-515.

[38] CO_2 forms carbonic acid when dissolved in water. Over time, the solid calcium carbonate ($CaCO_3$) on the seafloor will react with (neutralize) much of the carbonic acid that entered the oceans as CO_2 from the atmosphere. See David Archer et al., "Dynamics of Fossil Fuel CO_2 Neutralization by Marine $CaCO_3$," Global Biogeochemical Cycles, vol. 12, no. 2 (June 1998): pp. 259-276.

[39] One study, for example, suggests that the efficiency of the ocean sink has been declining at least since 2000; see Josep G. Canadell et al., "Contributions to Accelerating Atmospheric CO_2 Growth from Economic Activity, Carbon Intensity, and Efficiency of Natural Sinks," Proceedings of the National Academy of Sciences, vol. 104, no. 47 (Nov. 20, 2007), pp. 18866-18870.

[40] For more information on ocean acidification, see CRS Report R40143, Ocean Acidification, by Eugene H. Buck and Peter Folger.

[41] IPCC Special Report, p. 285.

[42] A CO_2 hydrate is a crystalline compound formed at high pressures and low temperatures by trapping CO_2 molecules in a cage of water molecules.

[43] K. Z. House, et al., "Permanent Carbon Dioxide Storage in Deep-Sea Sediments," Proceedings of the National Academy of Sciences, vol. 103, no. 33 (Aug. 15, 2006): pp. 12291-12295.

[44] P. G. Brewer, et al., "Deep Ocean Experiments with Fossil Fuel Carbon Dioxide: Creation and Sensing of a Controlled Plume at 4 km Depth," Journal of Marine Research, vol. 63, no. 1 (2005): p. 9-33.

[45] IPCC Special Report, p. 279.

[46] Ibid., p. 298.

[47] Virginia Gewin, "Ocean Carbon Study to Quit Hawaii," Nature, vol. 417 (June 27, 2002): p. 888.

[48] Jim Giles, "Norway Sinks Ocean Carbon Study," Nature, vol. 419 (Sept. 5, 2002): p. 6.

[49] Serpentine and olivine are silicate oxide minerals—combinations of the silica, oxygen, and magnesium—that react with CO_2 to form magnesium carbonates. Wollastonite, a silica oxide mineral containing calcium, reacts with CO_2 to form calcium carbonate (limestone). Magnesium and calcium carbonates are stable minerals over long time scales.

[50] Calera, a company based in Los Gatos, CA, has developed a process for mineral carbonation that it claims will sequester CO_2 and produce solid carbonate minerals that can be used in the manufacture of building materials. The Calera process is discussed in a CRS congressional distribution (CD) memorandum, available from Peter Folger at 7- 1517.

[51] IPCC Special Report, p. 40.

[52] Flood basalts are vast expanses of solidified lava, commonly containing olivine, that erupted over large regions in several locations around the globe. In addition to the Columbia River Plateau flood basalts, other well-known flood basalts include the Deccan Traps in India and the Siberian Traps in Russia.

[53] 2010 Carbon Sequestration Atlas, p. 30.

In: Carbon Capture and Sequestration   ISBN: 978-1-62257-810-8
Editors: A.C. Mitchell and R. Freeman   © 2013 Nova Science Publishers, Inc.

Chapter 2

CARBON CAPTURE AND SEQUESTRATION: RESEARCH, DEVELOPMENT, AND DEMONSTRATION AT THE U.S. DEPARTMENT OF ENERGY[*]

Peter Folger

SUMMARY

On March 27, 2012, the U.S. Environmental Protection Agency (EPA) proposed a new rule that would limit emissions to no more than 1,000 pounds of carbon dioxide (CO_2) per megawatt-hour of production from new fossil-fuel power plants with a capacity of 25 megawatts or larger.

EPA proposed the rule under Section 111 of the Clean Air Act. According to EPA, new natural gas-fired combined-cycle power plants should be able to meet the proposed standards without additional cost. However, new coal-fired plants would only be able to meet the standards by installing carbon capture and sequestration (CCS) technology.

The proposed rule has sparked increased scrutiny of the future of CCS as a viable technology for reducing CO_2 emissions from coal-fired power plants.

---

[*] This is an edited, reformatted and augmented version of a Congressional Research Service publication, CRS Report for Congress R42496, from www.crs.gov, prepared for Members and Committees of Congress, dated April 23, 2012.

The proposed rule also places a new focus on whether the U.S. Department of Energy's (DOE's) CCS research, development, and demonstration (RD&D) program will achieve its vision of developing an advanced CCS technology portfolio ready by 2020 for large-scale CCS deployment.

Congress has appropriated nearly $6 billion since FY2008 for CCS RD&D at DOE's Office of Fossil Energy: approximately $2.3 billion from annual appropriations and $3.4 billion from the American Recovery and Reinvestment Act (or Recovery Act).

The large and rapid influx of funding for industrial-scale CCS projects from the Recovery Act may accelerate development and deployment of CCS in the United States.

However, the future deployment of CCS may take a different course if the major components of the DOE program follow a path similar to DOE's flagship CCS demonstration project, FutureGen, which has experienced delays and multiple changes of scope and design since its inception in 2003. A question for Congress is whether FutureGen represents a unique case of a first mover in a complex, expensive, and technically challenging endeavor, or whether it indicates the likely path for all large CCS demonstration projects once they move past the planning stage.

Since enactment of the Recovery Act, DOE has shifted its RD&D emphasis to the demonstration phase of carbon capture technology. The shift appears to heed recommendations from many experts who called for large, industrial-scale carbon capture demonstration projects (e.g., 1 million tons of CO_2 captured per year). Funding from the Recovery Act for large-scale demonstration projects was 40% of the total amount of DOE funding for all CCS RD&D from FY2008 through FY2012.

To date, there are no commercial ventures in the United States that capture, transport, and inject industrial-scale quantities of CO2 solely for the purposes of carbon sequestration. However, CCS RD&D in 2012 is just now embarking on commercial-scale demonstration projects for CO2 capture, injection, and storage. The success of these projects will likely bear heavily on the future outlook for widespread deployment of CCS technologies as a strategy for preventing large quantities of CO_2 from reaching the atmosphere while U.S. power plants continue to burn fossil fuels, mainly coal.

Given the pending EPA rule, congressional interest in the future of coal as a domestic energy source appears directly linked to the future of CCS. In the short term, congressional support for building new coal-fired power plants could be expressed through legislative action to modify or block the proposed EPA rule. Alternatively, congressional oversight of the CCS RD&D program could help inform decisions about the level of support for the program and help Congress gauge whether it is on track to meet its goals.

INTRODUCTION

Carbon capture and sequestration (or storage)—known as CCS—is a physical process that involves capturing manmade carbon dioxide (CO_2) at its source and storing it before its release to the atmosphere.

CCS could reduce the amount of CO_2 emitted to the atmosphere while allowing the continued use of fossil fuels at power plants and other large, industrial facilities.

An integrated CCS system would include three main steps: (1) capturing CO_2 at its source and separating it from other gases; (2) purifying, compressing, and transporting the captured CO_2 to the sequestration site; and (3) injecting the CO_2 into subsurface geological reservoirs. Following its injection into a subsurface reservoir, the CO_2 would need to be monitored for leakage and to verify that it remains in the target geological reservoir. Once injection operations cease, a responsible party would need to take title to the injected CO_2 and ensure that it stays underground in perpetuity.

The U.S. Department of Energy (DOE) has pursued research and development of aspects of the three main steps leading to an integrated CCS system since 1997.[1] Congress has appropriated nearly $6 billion since FY2008 for CCS research, development, and demonstration (RD&D) at DOE's Office of Fossil Energy: approximately $2.3 billion in total annual appropriations, and $3.4 billion from the American Recovery and Reinvestment Act (P.L. 111-5, enacted February 17, 2009, hereinafter referred to as the Recovery Act).

The large and rapid influx of funding for industrial-scale CCS projects from the Recovery Act may accelerate development and demonstration of CCS in the United States, particularly if the RD&D pursued by DOE's CCS program achieves its goals as outlined in the department's 2010 RD&D *CCS Roadmap*.[2]

However, the future deployment of CCS may take a different course if the major components of the DOE program follow a path similar to DOE's FutureGen project, which has experienced delays and multiple changes of scope and design since its inception in 2003.

This report aims to provide a snapshot of the DOE CCS program, including its current funding levels and the budget request for FY2013, together with some discussion of the program's achievements and prospects for success in meeting its stated goals.

ISSUES FOR CONGRESS

EPA Proposed Rule Limiting CO_2 Emissions from Power Plants

On March 27, 2012, the U.S. Environmental Protection Agency (EPA) proposed a new rule that would limit emissions from new fossil-fuel power plants to no more than 1,000 pounds of CO_2 per megawatt-hour of energy produced. It would apply to plants with a generating capacity of greater than 25 megawatts.[3] EPA proposed the rule under Section 111 of the Clean Air Act, amending 40 C.F.R. Part 60. According to EPA, new natural gas-fired combined-cycle power plants should be able to meet the proposed standards without additional cost. However, new coal-fired plants would only be able to meet the standards by using CCS.[4]

The proposed rule has sparked increased scrutiny of the future of CCS as a viable technology for reducing CO_2 emissions from coal-fired power plants. The proposed rule also places a new focus on DOE's CCS RD&D program— whether it will achieve its vision of "having an advanced CCS technology portfolio ready by 2020 for large-scale CCS demonstration that provides for the safe, cost-effective carbon management that will meet our Nation's goals for reducing [greenhouse gas] emissions."[5]

On March 27, 2012, EPA Administrator Lisa Jackson reportedly stated that CCS would be commercially available within 10 years.[6] Further, EPA's regulatory impact analysis, which accompanied the proposed rule, stated that "EPA intends this rule to send a clear signal about the future of CCS technology that, in conjunction with other policies such as Department of Energy (DOE) financial assistance, the agency estimates will support development and demonstration of CCS technology from coal-fired plants at commercial scale, if that financial assistance is made available under appropriate market conditions."[7]

The prospects for building new coal-fired electricity generating plants depend on many factors, such as costs of competing fuel sources (e.g., natural gas), electricity demand, regulatory costs, infrastructure (including rail) and electric grid development, and others. However, the EPA proposed rule clearly identifies CCS as the essential technology required if new coal-fired power plants are to be built in the United States.[8]

Congress has appropriated funding for DOE to pursue CCS research and development since 1997 and recently acknowledged the importance of CCS technology by awarding $3.4 billion from the Recovery Act to CCS programs at DOE. Given the pending EPA rule, congressional interest in the future of coal as a domestic energy source appears directly linked to the future of CCS. In the short term, congressional support for building new coal-fired power plants could be expressed through legislative action to modify or block the proposed EPA rule. Alternatively, congressional oversight of the DOE CCS RD&D program could help inform decisions about the level of support for the program and help Congress gauge whether the program is on track to meet its goals. Despite Administrator Jackson's assurances that CCS will be commercially available in 10 years, the history of CCS RD&D at DOE and the pathway of some its signature programs, such as FutureGen, invite questions about whether the RD&D results will enable widespread deployment of CCS in the United States within the next decade.

Legislation

Although DOE has pursued aspects of CCS RD&D since 1997, the Energy Policy Act of 2005 (P.L. 109-58) provided a 10-year authorization for the basic framework of CCS research and development at DOE.[9] The Energy Independence and Security Act of 2007 (EISA, P.L. 110-140) amended the Energy Policy Act of 2005 to include, among other provisions, authorization for seven large-scale CCS demonstration projects (in addition to FutureGen) that would integrate the carbon capture, transportation, and sequestration steps.[10] (Large-scale demonstration programs and their potential significance are discussed below.) It can be argued that, since enactment of EISA, the focus and funding within the CCS RD&D program has shifted towards large-scale capture technology development through these and other demonstration projects.

In addition to the annual appropriations provided for CCS RD&D, the legislation most significant to federal CCS RD&D program activities since passage of EISA has been the Recovery Act (P.L. 111-5). As discussed below, $3.4 billion in funding from the Recovery Act was intended to expand and accelerate the commercial deployment of CCS technologies to allow for commercial-scale demonstration in both new and retrofitted power plants and industrial facilities by 2020.

In the 111th Congress, two bills that would have authorized a national cap-and-trade system for limiting the emission of greenhouse gases (H.R. 2454 and S. 1733) also would have created programs aimed at accelerating the commercial availability of CCS. The programs would have generated funding from a surcharge on electricity delivered from the combustion of fossil fuels—

approximately $1 billion per year—and made the funding available for grants, contracts, and financial assistance to eligible entities seeking to develop CCS technology. Another source of funding in the bills was to come from a program that would distribute emission allowances to "early movers," entities that installed CCS technology on up to a total of 20 gigawatts generating capacity. The House of Representatives passed H.R. 2454, but neither bill was enacted.

In the 112th Congress, a few bills have been introduced that would address aspects of CCS RD&D. The Department of Energy Carbon Capture and Sequestration Program Amendments Act of 2011 (S. 699) would provide federal indemnification of up to $10 billion per project to early adopters of CCS technology (large CCS demonstration projects).[11]

The New Manhattan Project for Energy Independence (H.R. 301) would create a system of grants and prizes for a variety of technologies, including CCS, that would contribute to reducing U.S. dependence on foreign sources of energy. Other bills introduced would provide tax incentives for the use of CO_2 in enhanced oil recovery (S. 1321), or would eliminate the minimum capture requirement for the CO_2 sequestration tax credit (H.R. 1023). Other bills were also introduced that would affect other aspects of CCS RD&D financing, such as loan guarantees. To date, none of the bills introduced in the 112th Congress affecting federal CCS RD&D, other than annual appropriations, has been enacted.

CCS RESEARCH, DEVELOPMENT, AND DEMONSTRATION: OVERALL GOALS

The U.S. Department of Energy states that the mission for the DOE Office of Fossil Energy is "to ensure the availability of ultra-clean (near-zero emissions), abundant, low-cost domestic energy from coal to fuel economic prosperity, strengthen energy security, and enhance environmental quality."[12] Over the past several years, the DOE Fossil Energy Research and Development Program has increasingly shifted activities performed under its Coal Program toward emphasizing CCS as the main focus.[13] The Coal Program itself represented 68% of total Fossil Energy Research and Development appropriations in FY2011 and 75% in FY2012,[14] indicating that CCS has come to dominate coal R&D at DOE. This reflects DOE's view that "there is a growing consensus that steps must be taken to significantly reduce [greenhouse gas] emissions from energy use throughout the world at a pace consistent to stabilize atmospheric concentrations of CO_2, and that CCS is a promising option for addressing this challenge."[15]

DOE also acknowledges that the cost of deploying currently available CCS technologies is very high, and that to be effective as a technology for mitigating greenhouse gas emissions from power plants, the costs for CCS must be reduced.[16] The challenge of reducing the costs of CCS technology is difficult to quantify, in part because there are no examples of currently operating commercial-scale coal-fired power plants equipped with CCS. Nor is it easy to predict when lower-cost CCS technology will be available for widespread deployment in the United States. Nevertheless, DOE observes that "the United States can no longer afford the luxury of conventional long-lead times for RD&D to bear results."[17] Thus the coal RD&D program is focused on achieving results that would allow for an advanced CCS technology portfolio to be ready by 2020 for large-scale demonstration.

The following section describes the components of the CCS activities within DOE's coal R&D program and their funding history over the past five fiscal years. This report focuses on this time period because during that time the Recovery Act was enacted and was expected to accelerate the transition of CCS technology to industry for deployment and commercialization.[18]

Program Areas

The 2010 RD&D *CCS Roadmap* described 10 different program areas pursued by DOE's Coal Program within the Office of Fossil Energy: (1) Innovations for Existing Plants (IEP); (2) Advanced Integrated Gasification Combined Cycle (IGCC); (3) Advanced Turbines; (4) Carbon Sequestration; (5) Solid State Energy Conversion Fuel Cells; (6) Fuels; (7) Advanced Research; (8) Clean Coal Power Initiative (CCPI); (9) FutureGen; and (10) Industrial Carbon Capture and Storage Projects (ICCS).[19] *Table 1* shows funding for these program areas through FY2010, including funding provided by the Recovery Act, and for two subprogram areas: (1) Site Characterization, Training, Program Direction; and (2) Carbon Sequestration (Focus Area for CCS Research). DOE changed the program structure after FY2010. *Table 1* reflects those changes and attempts to show how the new program structure matches the old program structure after FY2010.

Some program areas are directly focused on one or more of the three steps of CCS: capture, transportation, and storage. For example, the Carbon

Sequestration program area focuses on the third step: evaluating prospective sites for long-term storage of CO_2 underground. In contrast, FutureGen from the outset was envisioned as combining all three steps: a zero-emission fossil fuel plant that would capture its emissions and sequester them in a geologic reservoir.

Other program areas include research that, at first glance, does not seem directly applicable to the three steps of CCS. For example, the CCPI and IEP programs include research aimed at achieving gains in power plant efficiency. These technologies would presumably reduce CO_2 emissions per unit of power generated, even if they were not directly integrated into a carbon capture and sequestration scheme.

The DOE *CCS Roadmap* notes that each program area, however, contributes to the overall CCS RD&D effort.

As outlined in the 2010 *CCS Roadmap*, RD&D efforts in several of these program areas overlap. For example, the lead role for pre-combustion carbon capture technology RD&D falls under the Carbon Sequestration program area, but additional pre-combustion technology RD&D also occurs under the IGCC program area and under the Fuels program area.[20] Post-combustion and oxy-combustion carbon capture technology RD&D is conducted in a different program area: IEP, which is focused on technology for new and existing pulverized coal power plants.

In addition, the Advanced Research program area conducts research on new breakthrough carbon capture technologies, as does the National Energy Technology Laboratory (NETL) Office of Research and Development (described as external and internal research, respectively, in the 2010 *CCS Roadmap*).

Thus, at least six different program areas identified in the 2010 *CCS Roadmap* conduct research on carbon capture technology RD&D.

RD&D is also divided among different industrial sectors in two program areas: the Clean Coal Power Initiative (CCPI) and Industrial Carbon Capture and Storage Projects (ICCS). The CCPI program area focuses on the demonstration phase of carbon capture technology for coal-based power plants.

The ICCS area demonstrates carbon capture technology for the non-power plant industrial sector.[21] Both these program areas focus on the *demonstration* component of RD&D, and account for $2.3 billion of the $3.4 billion appropriated for CCS RD&D in the Recovery Act in FY2009. From the budgetary perspective, the Recovery Act funding shifted the emphasis of CCS RD&D to large, industrial demonstration projects for carbon capture.

Recovery Act funding for CCPI and ICCS projects was 40% of the *total* amount of funding for all CCS RD&D from FY2008 through FY2012 (*Table 1*).

If funding for FutureGen—which is arguably a large CCS demonstration project—were added, then funding for demonstration projects would comprise approximately 58% of the total amount allocated for CCS RD&D over the past five years.

This shift in emphasis to the demonstration phase of carbon capture technology is not surprising, and appears to heed recommendations from many experts who called for large, industrial-scale carbon capture demonstration projects.[22]

Primarily, the call for large-scale CCS demonstration projects that capture 1 million metric tons or more of CO_2 per year reflects the need to reduce the additional costs to the power plant or industrial facility associated with capturing the CO_2 before it is emitted to the atmosphere. The capture component of CCS is the costliest component, according to most experts.[23] The higher costs of power plants with CCS, compared to plants without CCS, and the uncertainty in cost estimates results in part from a dearth of information about outstanding technical questions in carbon capture technology at the industrial scale.[24]

In comparative studies of cost estimates for other environmental technologies, such as for power plant scrubbers that remove sulfur and nitrogen compounds from power plant emissions (SO_2 and NOx), some experts note that the farther away a technology is from commercial reality, the more uncertain is its estimated cost.

At the beginning of the RD&D process, initial cost estimates could be low, but could typically increase through the demonstration phase before decreasing after successful deployment and commercialization. *Figure 1* shows a cost estimate curve of this type.

Deploying commercial-scale CCS demonstration projects—an emphasis within the DOE CCS RD&D program—would therefore provide cost estimates closer to operational conditions rather than laboratory- or pilot-plant-scale projects. In the case of SO_2 and NOx scrubbers, efforts typically took two decades or more to bring new concepts (such as combined SO_2 and NOx capture systems) to the commercial stage. As *Figure 1* indicates, costs for new technologies tend to fall over time with successful deployment and commercialization. It would be reasonable to expect a similar trend for CO_2 capture costs if the technologies become widely deployed.[25]

Table 1. DOE Carbon Capture and Storage Research, Development, and Demonstration Program Areas (funding in $ thousands, FY2008-FY2013)

Program	FY2008	FY2009	Recovery Act	FY2010	Restructured Program after FY2010[a]	FY2011	FY2012 (enacted)	Totals (FY2008-FY2012)	FY2013 (request)
FutureGen	72,262	0	1,000,000	0	FutureGen 2.0	0	0	1,072,262	0
Clean Coal Power Initiative (CCPI)	67,444	288,174	800,000	0	CCS Demonstrations	0	0	1,155,618	0
Industrial Carbon Capture and Storage Projects			1,520,000	0		0	0	1,520,000	0
Site Characterization, Training, Program Direction			80,000	0		0	0	80,000	0
Innovations for Existing Plants (IEP)	35,083	48,600	—	50,630	Carbon Capture Research	58,703	68,898	261,914	60,438
Advanced Integrated Gasification Combined Cycle (IGCC)	52,029	63,409	—	61,341	Advanced Energy Systems (IGCC)	78,338	54,942	310,059	42,604
Advanced Turbines (Hydrogen Turbines)	23,125	27,216	—	31,158	Advanced Energy Systems (Hydrogen Turbines)	30,106	15,000	126,605	12,589
Carbon Sequestration (Greenhouse Gas Control)	105,985	132,192	80,000	136,313	Carbon Storage Research	111,195	105,684	591,369	85,751
Carbon	9,635	13,608	—	13,631	Carbon Storage	9,717	9,726	56,317	9,726

Program	FY2008	FY2009	Recovery Act	FY2010	Restructured Program after FY2010[a]	FY2011	FY2012 (enacted)	Totals (FY2008-FY2012)	FY2013 (request)
Sequestration (Focus Area for CCS Research)					(Focus Area for CCS Research)				
Fuels (Hydrogen from Coal/Coal to Liquids)	24,088	24,300	—	24,341	Advanced Energy Systems (Hydrogen from Coal/Coal to Liquids)	11,661	5,000	89,390	0
Solid State Energy Conversion Fuel Cells	53,956	56,376	—	48,683	Advanced Energy Systems (Fuel Cells)	48,522	25,000	232,537	0
Advanced Research	36,264	27,389	—	27,388	Cross Cutting Research	41,446	49,134	181,621	29,750
					NETL Coal Research and Development	0	35,011	35,011	35,011
Totals	479,871	681,264	3,400,000	393,485		389,688	368,395	5,712,703	275,869

Source: U.S. Department of Energy, FY2013 Congressional Budget Request, volume 3, Fossil Energy Research and Development, http://www.cfo.doe.gov/budget/13budget/Content/Volume3.pdf; U.S. Department of Energy, Carbon Sequestration, Recovery Act, http://www.fe.doe.gov/recovery/index.html.

Notes: Advanced Energy Systems excludes funding for Hydrogen from Coal, Coal and Coal-Biomass to Liquids, and Solid Oxide Fuel Cells programs.

a. In FY2010, post-combustion carbon capture research was included in the Carbon Sequestration (Greenhouse Gas Control) line item, to reflect the transition to a new budget structure proposed for FY2012. This table attempts to show the broad program areas for CCS RD&D before and after the DOE changed the budget structure after FY2010.

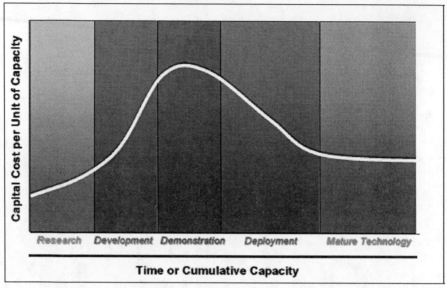

Source: Adapted from S. Dalton, "CO2 Capture at Coal Fired Power Plants—Status and Outlook," 9thInternational Conference on Greenhouse Gas Control Technologies, Washington, DC, November, 16-20, 2008.

Figure 1. Typical Trend in Cost Estimates for a New Technology As It Develops from a Research Concept to Commercial Maturity.

RECOVERY ACT FUNDING FOR CCS PROJECTS: A LYNCHPIN FOR SUCCESS?

The bulk of Recovery Act funds for CCS ($3.32 billion, or 98%) was directed to three programs: Clean Coal Power Initiative; Industrial Carbon Capture and Storage Projects, and FutureGen (*Table 1*). Under the 2010 *CCS Roadmap*, and with the large infusion of funding from the Recovery Act, DOE's goal is to develop the technologies to allow for commercial-scale demonstration in both new and retrofitted power plants and industrial facilities by 2020. The DOE 2011 *Strategic Plan* sets a more specific target: bring at least five commercial-scale CCS demonstration projects online by 2016.[26]

It could be argued that in its allocation of Recovery Act funding, DOE was heeding the recommendations of experts (see footnote 22) who identified commercial-scale demonstration projects as the most important component, the lynchpin, for future development and deployment of CCS in the United States. It could also be argued that much of the future success of CCS is riding on these three programs. Accordingly, the following section provides a snapshot of the CCPI, ICCS, and FutureGen programs, and a brief discussion of some of their accomplishments and challenges.

Clean Coal Power Initiative

The CCPI was an ongoing program prior to the $800 million funding increase from the Recovery Act (see *Table 1*). Recovery Act funding now is being used to expand activities in CCPI Round 3 beyond developing technologies to reduce sulfur, nitrogen, and mercury pollutants from power plants.[27] After enactment of the Recovery Act, DOE did not request additional funding for CCPI under its Fossil Energy program in the annual appropriations process (*Table 1* shows zeroes in the columns for FY2010-FY2013). Rather, in the FY2010 DOE budget justification, DOE stated that funding for the CCPI demonstration projects in Round 3 would be supported through the Recovery Act, and as a result "DOE will make dramatic progress in demonstrating CCS at commercial scale using these funds without the need for additional resources for demonstration in 2010."[28]

According to the 2010 DOE *CCS Roadmap*, Recovery Act funds are being used for CCPI demonstration projects to "allow researchers broader CCS commercial-scale experience by expanding the range of technologies, applications, fuels, and geologic formations that are being tested."[29] DOE selected six projects under CCPI Round 3 through two separate solicitations.[30] The total DOE share of funding would have been $1.54 billion for the six projects in five states: Texas, California, North Dakota, West Virginia, and Alabama (*Table 2*). However, the projects in Alabama, North Dakota, and West Virginia withdrew from the program, and currently the DOE share for the remaining three projects is $881 million (of a total of over $6 billion for total expected costs). With the withdrawal of three CCPI Round 3 projects, DOE's share of the total program costs shrank from nearly 20% to approximately 13%.

Reasons for Withdrawal from the CCPI Program

Southern Company—Plant Barry 160 MW Project

Southern Company withdrew its Alabama Plant Barry project from the CCPI program on February 22, 2010, slightly more than two months after

DOE Secretary Chu announced $295 million in DOE funding for the 11-year, $665 million project that would have captured up to 1 million tons of CO_2 per year from a 160 megawatt coal-fired generation unit.[31] According to some sources, Southern Company's decision was based on concern about the size of the company's needed commitment (approximately $350 million) to the project, and its need for more time to perform due diligence on its financial commitment, among other reasons.[32]

Table 2. DOE CCPI Round 3 Projects

CCPI Round 3 Project	Location	DOE Share of Funding ($ millions)	Total Project Cost ($ millions)	Percent DOE Share	Metric Tons of CO_2 Captured Annually (millions)	Project Status
Texas Clean Energy Project	Penwell, TX	450	1,727	26%	2.7	Active
Hydrogen Energy California Project	Kern County, CA	408	4,008	10%	1.8 or 2.5	Active
NRG Energy Project	Thompsons, TX	167	334	50%	0.4	Active
AEP Mountaineer Project	New Haven, WV	334	668	50%	1.5	Withdrawn
Southern Company Project	Mobile, AL	295	665	44%	1	Withdrawn
Basin Electric Power Project	Beulah, ND	100	387	26%	0.9	Withdrawn
Total		1,541	7,789	19.8%	8.3	
Total, Active Projects[a]		812	6,069	13.4%	4.9	

Source: DOE Fossil Energy Techline; Environment News Service (March 12, 2010), http://www.ens-newswire.com/ ens/mar2010/2010-03-12-093.html; NETL CCPI website, http://www.netl.doe.gov/technologies/coalpower/cctc/ccpi/ index.html.

Notes: DOE funding for the NRG Energy Project was initially announced as up to $154 million (see March 9, 2009, DOE Techline, http://www.fossil.energy.gov /news/techlines/2010/10005- NRG_Energy_Selected_to_Receive_DOE.html). A May 2010 DOE fact sheet indicates that funding for NRG is $167 million (http://www.netl.doe.gov/ publications/factsheets/project/FE0003311.pdf).

[a] Total include amounts that were reallocated from withdrawn projects to active projects.

Southern Company continues work on a much smaller CCS project that would capture CO_2 from a 25 MW unit at Plant Barry.

Basin Electric Power—Antelope Valley 120 MW Project

On July 1, 2009, Secretary Chu announced $100 million in DOE funding for a project that would capture approximately 1 million tons of CO_2 per year from a 120 MW electric-equivalent gas stream from the Antelope Valley power station near Beulah, ND.[33] In December 2010, the Basin Electric Power Cooperative withdrew its project from the CCPI program, citing both regulatory uncertainty with regard to capturing CO_2 and uncertainty about the project's cost (one source indicates that the company estimated $500 million total cost; DOE estimated $387 million—see *Table 2*).[34] The project would have supplied the captured CO_2 to an existing pipeline that transports CO_2 from the Great Plains Synfuels Plant near Beulah for enhanced oil recovery in Canada's Weyburn field approximately 200 miles north in Saskatchewan.

American Electric Power—Mountaineer 235 MW Project

The most recent cancellation was announced in July 2011, when American Electric Power decided to halt its plans to build a carbon capture plant for a 235 MW generation unit at its 1.3 gigawatt Mountaineer power plant in New Haven, WV. The project represented Phase 2 of an ongoing CCPI project. Secretary Chu had earlier announced a $334 million award for the project on December 4, 2009.[35] According to some sources, AEP dropped the project because the company was not certain that state regulators would allow it to recover the additional costs for the CCS project through rate increases charged to its customers.[36] In addition, company officials cited broader economic and policy conditions as reasons for cancelling the project.[37] Some commentators suggested that congressional inaction on setting limits on greenhouse gas emissions, as well as the weak economy, may have diminished the incentives for a company like AEP to invest in CCS.[38] One source concluded that "Phase 2 has been cancelled due to unknown climate policy."[39]

Reshuffling of Funding for CCPI

According to DOE, $140 million of the $295 million previously allotted to the Southern Company Plant Barry project was divided between the Texas Clean Energy project and the Hydrogen Energy California project, together with additional DOE funding, so that each project received an additional $100 million total above its initial awards.[40] The remaining funding from the canceled Plant Barry project (up to $154 million) was allotted to the NRG Energy project in Texas (see *Table 2*).[41]

According to a DOE source, selection of the Basin Electric Power project was announced but a cooperative agreement was never awarded by DOE.[42] Funds that were to be obligated for the Basin project could therefore have been reallocated within the department, but were rescinded by Congress in FY2011 appropriations.

Some of the funding for the AEP Mountaineer project was rescinded by Congress in FY2012 appropriations legislation (P.L. 112-74). In the report accompanying P.L. 112-74, Congress rescinded a total of $187 million of prior-year balances from the Fossil Energy Research and Development account.[43] The rescission did not apply to amounts previously appropriated under P.L. 111-5; however, funding for the AEP Mountaineer project that was provided by the Recovery Act and not spent was returned to the Treasury and not made available to the CCPI program.[44]

Industrial Carbon Capture and Storage Projects

The original DOE ICCS program was divided into two main areas: Area 1, consisting of large industrial demonstration projects; and Area 2, consisting of projects to test innovative concepts for the beneficial reuse of CO_2.[45] Under Area 1, the first phase of the program consisted of 12 projects cost-shared with private industry, intended to increase investment in clean industrial technologies and sequestration projects. Phase I projects averaged approximately seven months in duration. Following Phase I, DOE selected three projects for Phase 2 for design, construction, and operation.[46] The three Phase 2 projects are listed as large-scale demonstration projects in *Table 3*. The total share of DOE funding for the three projects, provided by the Recovery Act, is $686 million, or approximately 64% of the sum total Area 1 program cost of $1.075 billion.

Under Area 2, the initial phase consisted of $17.4 million in Recovery Act funding and $7.7 million in private-sector funding for 12 projects to engage in feasibility studies to examine the beneficial reuse of CO_2.[47] In July 2010, DOE selected 6 projects from the original 12 projects for a second phase of funding to find ways of converting captured CO_2 into useful products such as fuel, plastics, cement, and fertilizer. The 6 projects are listed under "Innovative Concepts/Beneficial Use" in *Table 3*.

Table 3. DOE Industrial Carbon Capture and Storage (ICCS) Projects
(showing DOE share of funding and total project cost)

ICCS Project Name	Location	Type of Project	DOE Share of Funding ($ millions)	Total Project Cost ($ millions)	Percent DOE Share
Air Products & Chemicals, Inc.	Port Arthur, TX	Large-Scale Demonstration	284	431	66%
Archer Daniels Midland Co.	Decatur, IL	Large-Scale Demonstration	141	208	68%
Leucadia Energy, LLC	Lake Charles, LA	Large-Scale Demonstration	261	436	60%
Alcoa, Inc.	Alcoa Center, PA	Innovative Concepts/Beneficial Use	13.5	16.9	80%
Novomer, Inc.	Ithaca, NY	Innovative Concepts/Beneficial Use	20.5	25.6	80%
Touchstone Research Lab, Ltd.	Triadelphia, PA	Innovative Concepts/Beneficial Use	6.7	8.4	80%
Phycal, LLC	Highland Heights, OH	Innovative Concepts/Beneficial Use	51.4	65	80%
Skyonic Corp.	Austin, TX	Innovative Concepts/Beneficial Use	28	39.6	70%
Calera Corp.	Los Gatos, CA	Innovative Concepts/Beneficial Use	21.4	42.7	50%
Air Products & Chemicals, Inc.	Allentown, PA	Advanced Gasification Technologies	71.7	75	96%
Eltron Research & Development, Inc.	Boulder, CO	Advanced Gasification Technologies	71.4	73.7	97%
Research Triangle Institute	Research Triangle Park, NC	Advanced Gasification Technologies	168.8	174	97%
GE Energy	Schenectady, NY	Advanced Turbo-Machinery	31.3	62.6	50%

Table 3. (Continued)

ICCS Project Name	Location	Type of Project	DOE Share of Funding ($ millions)	Total Project Cost ($ millions)	Percent DOE Share
Siemens Energy	Orlando, FL	Advanced Turbo-Machinery	32.3	64.7	50%
Clean Energy Systems, Inc.	Rancho Cordova, CA	Advanced Turbo-Machinery	30	42.9	70%
Ramgen Power Systems	Bellevue, WA	Advanced Turbo-Machinery	50	79.7	63%
ADA-ES, Inc.	Littleton, CO	Post-Combustion Capture	15	18.8	80%
Alstom Power	Windsor, CT	Post-Combustion Capture	10	12.5	80%
Membrane Technology & Research, Inc	Menlo Park, CA	Post-Combustion Capture	15	18.8	80%
Praxair	Tonawanda, NY	Post-Combustion Capture	35	55.6	63%
Siemens Energy, Inc.	Pittsburgh, PA	Post-Combustion Capture	15	18.8	80%
Board of Trustees U. of IL	Champaign, IL	Geologic Site Characterization	5	6.5	77%
N. American Power Group, Ltd.	Greenwood Village, CO	Geologic Site Characterization	5	7.85	64%
Sandia Technologies, LLC	Houston, TX	Geologic Site Characterization	4.38	5.63	78%
S. Carolina Research Foundation	Columbia, SC	Geologic Site Characterization	5	6.25	80%
Terralog Technologies USA, Inc.	Arcadia, CA	Geologic Site Characterization	5	6.25	80%
U. of Alabama	Tuscaloosa, AL	Geologic Site Characterization	5	10.8	46%
U. of Kansas Center for Research, Inc	Lawrence, KS	Geologic Site Characterization	5	6.29	80%
U. of Texas at	Austin, TX	Geologic Site	5	6.25	80%

ICCS Project Name	Location	Type of Project	DOE Share of Funding ($ millions)	Total Project Cost ($ millions)	Percent DOE Share
Austin		Characterization			
U. of Utah	Salt Lake City, UT	Geologic Site Characterization	5	7.23	69%
U. of Wyoming	Laramie, WY	Geologic Site Characterization	5	5	100%
		Totals	1,422.4	2,038.4	70%

Source: Emails from Regis K. Conrad, Director, Division of Cross-Cutting Research, DOE, March 20 and March 27, 2012; U.S. DOE, National Energy Technology Laboratory, Major Demonstrations, Industrial Capture and Storage (ICCS): Area 1, http://www.netl.doe.gov/technologies/coalpower/cctc/iccs1/index.html#; U.S. DOE, Carbon Capture and Storage from Industrial Sources, Industrial Carbon Capture Project Selections, http://fossil.energy projects/iccs_ projects_ 0907101.pdf.

Notes: Table is ordered from top to bottom by type of project: Large-Scale Demonstration; Innovative Concepts/Beneficial Use; Advanced Gasification Technologies; Advanced Turbo-Machinery; Post-Combustion Capture; and Geologic Site Characterization. Totals may not add due to rounding.

The total share of DOE funding for the 6 projects, provided by the Recovery Act, is $141.5 million, or approximately 71% of the sum total cost of $198.2 million.

Since its original conception, the DOE ICCS program has expanded with an additional 22 projects, funded under the Recovery Act, to accelerate promising technologies for CCS.[48]

In its listing of the 22 projects, DOE groups them into four general categories: (1) Large-Scale Testing of Advanced Gasification Technologies; (2) Advanced Turbo-Machinery to Lower Emissions from Industrial Sources; (3) Post-Combustion CO_2 Capture with Increased Efficiencies and Decreased Costs; and (4) Geologic Storage Site Characterization.[49]

The total share of DOE funding for the 22 projects, provided by Recovery Act, is $594.9 million, or approximately 78% of the sum total cost of $765.2 million.

Overall, the total share of federal funding for all the ICCS projects combined is $1.422 billion, or approximately 70% of the sum total cost of $2.038 billion.

FutureGen—A Special Case?

Brief History Since 2003

On February 27, 2003, President Bush proposed a 10-year, $1 billion project to build a coal-fired power plant that would integrate carbon sequestration and hydrogen production while producing 275 megawatts of electricity, enough to power about 150,000 average U.S. homes. As originally conceived, the plant would have been a coal-gasification facility and would have produced and sequestered between 1 million and 2 million tons of CO_2 annually. On January 30, 2008, DOE announced that it was "restructuring" the FutureGen program away from a single, state-of-the-art "living laboratory" of integrated R&D technologies—a single plant—to instead pursue a new strategy of multiple commercial demonstration projects.[50] In the restructured program, DOE would support up to two or three demonstration projects of at least 300 megawatts that would sequester at least 1 million tons of CO_2 per year.

In the Bush Administration's FY2009 budget, DOE requested $156 million for the restructured FutureGen program, and specified that the federal cost-share would only cover the CCS portions of the demonstration projects, not the entire power system. However, after the Recovery Act was enacted on February 17, 2009, Secretary Chu announced an agreement with the FutureGen Alliance—an industry consortium—to advance construction of the FutureGen plant built in Mattoon, IL, the site selected by the FutureGen Alliance in 2007.[51] Further, DOE anticipated that $1 billion of funding from the Recovery Act would be used to support the project.[52]

On August 5, 2010, Secretary Chu announced the $1 billion award, from Recovery Act funds, to the FutureGen Alliance, Ameren Energy Resources, Babcock & Wilcox, and Air Liquide Process & Construction, Inc., to build FutureGen 2.0.[53] FutureGen 2.0 differs from the original concept for the plant, because it would retrofit Ameren's existing power plant in Meridosia, IL, with oxycombustion technology at a 202 MW, oil-fired unit,[54] rather than build a new state-of-the-art plant in Mattoon. DOE announced that Mattoon would serve as the storage site for the CO_2 captured in Meridosia, and the two sites would be connected by a CO_2 pipeline to transport the gas.

The choice of Mattoon as the storage site met with opposition from the community in Coles County, IL, which includes the town of Mattoon. A group representing business development in Coles County sent a letter to Senator Dick Durbin (IL) stating that the community did not wish to be part of the new FutureGen 2.0 project with the Mattoon site solely as a storage facility.[55]

Subsequently, DOE and the FutureGen Alliance announced details of the process for selecting a new storage site in Illinois that would receive and store the 1 million or more metric tons of CO_2 from the Meridosia plant, and possibly CO_2 from other sources.[56] On February 28, 2011, the FutureGen Alliance announced that it had selected Morgan County, IL, as the location for the storage site.[57] Morgan County is west of Springfield, IL, and north of St. Louis, MO. Meridosia Unit 4, which would be repowered and retrofitted with a coal oxy-combustion carbon capture unit under FutureGen 2.0, is located about 20 miles west of Springfield.[58]

Current Status

The FutureGen 2.0 project has two parts: oxy-combustion carbon capture technology, implemented through a cooperative agreement with Ameren Energy Resources and industrial partners; and a pipeline and regional CO_2 storage reservoir project, implemented through a cooperative agreement with the FutureGen Alliance.[59] In addition to the $1 billion provided under the Recovery Act for the overall FutureGen 2.0 project, $53.6 million of prior appropriations would be provided for the pipeline and regional CO_2 storage portion. For the oxy-combustion carbon capture technology portion, DOE would provide $589.7 million, or 80%, of the $737 million in total costs. For the pipeline and regional CO_2 storage portion, DOE would provide $458.6 million, or 83%, of the $553 million in total costs.[60] The total federal and private-sector costs for FutureGen 2.0 are estimated to be almost $1.3 billion. The federal funding was formally awarded on October 1, 2010, and the award ends on December 31, 2020.

Challenges to FutureGen—A Similar Path for Other Demonstration Projects?

Slightly over a year after Secretary Chu announced the Recovery Act-funded award for FutureGen 2.0, Ameren Energy announced it was closing the Meridosia power plant, citing additional costs imposed by the requirement for environmental controls mandated by EPA under the authority of the Clean Air Act.[61] Ameren initially stated that the closure would not affect plans for FutureGen 2.0; however, it later reportedly admitted that it could not participate in the project.[62] The FutureGen Alliance subsequently entered into negotiations seeking to purchase parts of the Meridosia power plant from Ameren, which would likely allow the project to move forward and bring the generating unit retrofitted with oxy-combustion technology into production in 2016.[63]

In addition to the challenges of securing private-sector financing, acquiring the generating unit from Ameren, retrofitting it with oxy-combustion technology to capture 90% of the CO_2 emitted, and building a pipeline to a storage site, the project faces other potential hurdles. The FutureGen Alliance will need to secure approval for a specific storage site in Morgan County, which would also involve acquiring rights from private landowners to store the captured CO_2 in underground pore space, in addition to leasing the surface to build the infrastructure for injecting the CO_2 underground. As those decisions draw closer, past experience suggests that some local opposition to injecting and storing CO_2 indefinitely in the subsurface could emerge, and could derail the project unless a consensus within the local community is achieved.[64]

A question for Congress is whether FutureGen represents a unique case of a first mover in a complex, expensive, and technically challenging endeavor, or whether it represents all large CCS demonstration projects once they move past the planning stage. As discussed above, approximately $3.3 billion of Recovery Act funding is committed to large demonstration projects, including FutureGen. A rationale for committing such a substantial level of funding to demonstration projects was to scale up CCS RD&D more quickly than had been the pace of technology development prior to enactment of the Recovery Act. However, if all the CCS demonstration projects encounter similar changes in scope, design, location, and cost as FutureGen, the chances of meeting goals laid out in the DOE 2010 *Strategic Plan*—namely, to bring at least five commercial-scale CCS demonstration projects online by 2016—may be in jeopardy.

Alternatively, one could argue that FutureGen from its original conception was unique. None of the other large-scale demonstration projects share the same original ambitious vision: to create a new, one-of-a-kind, CCS plant from the ground up.

Even though the individual components of FutureGen—as originally conceived—were not themselves new innovations, combining the capture, transportation, and storage components together into a 250-megawatt functioning power plant could be considered unprecedented and therefore most likely to experience delays at each step in development.

Scholars have described the stages of technological change in different schemes, such as

- invention, innovation, adoption, diffusion;[65] or
- technology readiness levels (TRLs) ranging from TRL 1 (basic technology research) to TRL 9 (system test, launch, and operations);[66] or
- conceptual design, laboratory/bench scale, pilot plant scale, full-scale demonstration plant, and commercial process.[67]

FutureGen is difficult to categorize within these schemes, in part because the project spanned a range of technology development levels irrespective of the particular scheme. The original conception of the FutureGen project arguably had aspects of conceptual design through commercial processes—all five components of the scheme listed as the third bullet above—which meant that the project was intended to march through all stages in a linear fashion. As some scholars have noted, however, the stages of technological change are highly interactive, requiring learning by doing and learning by using, once the project progresses past its innovative stage into larger-scale demonstration and deployment.[68] The task of tackling all the stages of technology development in one project—the original FutureGen—might have been too daunting and, in addition to other factors, contributed to the project's erratic progress since 2003. It remains to be seen whether the current large-scale demonstration projects funded by DOE under CCPI Round 3 follow the path of FutureGen or instead achieve their technological development goals on time and within their current budgets.[69]

GEOLOGIC SEQUESTRATION/STORAGE: DOE RD&D FOR THE LAST STEP IN CCS

DOE has allocated between $115 million and $150 million per year for its carbon sequestration/ storage activities from FY2008 through FY2012, shown in Table 1 as "Carbon Sequestration (Greenhouse Gas Control)" and "Carbon Sequestration (Focus Area for CCS Research)" for FY2008 through FY2010; and as "Carbon Storage" after FY2010. The total amount allocated to carbon sequestration/carbon storage activities over that five-year period is slightly less than $650 million. In contrast with the carbon capture technology RD&D, which received nearly all of the $3.4 billion from Recovery Act funding, carbon sequestration/carbon storage activities received approximately $50 million in Recovery Act funds, or about 7% of the total funding amount for carbon sequestration since FY2008. Recovery Act funds were awarded for 10 projects to conduct site characterization of promising geologic formations for CO_2 storage.[70]

Brief History of DOE Geological Sequestration/Storage Activities

DOE has devoted the bulk of its funding for geological sequestration/ storage activities to RD&D efforts for injecting CO_2 into subsurface geological reservoirs. Injection and storage is the third step in the CCS process following the CO_2 capture step and CO_2 transport step. One part of the RD&D effort is characterizing geologic reservoirs (which received a $50 million boost from Recovery Act funds, as noted above); however, the overall program is much broader than just characterization, and has now reached the beginning of the phase of large-volume CO_2 injection demonstration projects across the country. According to DOE, these large-volume tests are needed to validate long-term storage in a variety of different storage formations of different depositional environments, including deep saline reservoirs, depleted oil and gas reservoirs, low permeability reservoirs, coal seams, shale, and basalt.[71] The large-volume tests can be considered injection experiments conducted at a commercial scale (i.e., approximately 1 million tons of CO_2 injected per year) that should provide crucial information on the suitability of different geologic reservoirs; monitoring, verification, and accounting of injected CO_2; risk assessment protocols for long-term injection and storage; and other critical challenges.

In 2003 DOE created seven regional carbon sequestration partnerships (RCSPs), essentially consortia of public and private sector organizations grouped by geographic region across the United States and parts of Canada.[72] The geographic representation was intended to match regional differences in fossil fuel use and geologic reservoir potential for CO_2 storage.[73] The RCSPs cover 43 states and 4 Canadian provinces and include over 400 organizations, according to the DOE 2010 *Strategic Plan*. *Table 4* shows the seven partnerships, the lead organization for each, and the states and provinces included. Several states belong to more than one RCSP. The RCSPs have pursued their objectives through three phases beginning in 2003:

(1) Characterization Phase (2003 to 2005), an initial examination of the region's potential for geological sequestration of CO_2; (2) Validation Phase (2005 to 2011), small-scale injection field tests (less than 500,000 tons of CO_2)

to develop a better understanding of how different geologic formations would handle large amounts of injected CO_2; and (3) Development Phase (2008 to 2018 and beyond), injection tests of at least 1 million tons of CO_2 to simulate commercial-scale quantities of injected CO_2.[74] The last phase is intended also to collect enough information to help understand the regulatory, economic, liability, ownership, and public outreach requirements for commercial deployment of CCS.

Table 4. Regional Carbon Sequestration Partnerships

Regional Carbon Sequestration Partnership (RCSP)	Lead Organization	States and Provinces in the Partnership
Big Sky Carbon Sequestration Partnership (BSCSP)	Montana State University-Bozeman	MT, WY, ID, SD, eastern WA, eastern OR
Midwest Geological Sequestration Consortium (MGSC)	Illinois State Geological Survey	IL, IN, KY
Midwest Regional Carbon Sequestration Partnership (MRCSP)	Battelle Memorial Institute	IN, KY, MD, MI, NJ, NY, OH, PA, WV,
Plains CO2 Reduction Partnership (PCOR)	University of North Dakota Energy and Environmental Research Center	MT, northeast WY, ND, SD, NE, MN, IA, MO, WI, Manitoba, Alberta, Saskatchewan, British Columbia (Canada)
Southeast Regional Carbon Sequestration Partnership (SECARB)	Southern States Energy Board	AL, AS, FL, GA, LA, MS, NC, SC, TN, TX, VA, portions of KY and WV
Southwest Regional Partnership on Carbon Sequestration (SWP)	New Mexico Institute of Mining and Technology	AZ, CO, OK, NM, UT, KS, NV, TX, WY
West Coast Regional Carbon Sequestration Partnership(WESTCARB)	California Energy Commission	AK, AZ, CA, HI, OR, NV, WA, British Columbia (Canada)

Source: DOE National Energy Technology Laboratory, Carbon Sequestration Regional Carbon Sequestration Partnerships, http://www.netl.doe.gov/technologies /carbon_seq/infrastructure/rcsp.html.

There are RD&D activities funded by DOE under its carbon sequestration/carbon storage program activities other than the RCSPs, such as geological storage technologies; monitoring, verification, and assessment; carbon use and reuse; and others. However, the RCSPs have averaged 50% or more of annual spending on carbon sequestration/carbon storage over the past three years. In the Administration's FY2013 budget request, funding for RCSPs would comprise 70% of the carbon storage research account. The RCSPs provide the framework and infrastructure for a wide variety of DOE geologic sequestration/storage activities.

Current Status and Challenges to Carbon Sequestration/Storage

The third phase—Development—is currently underway for all the RCSPs, and large-scale CO_2 injection has begun for the SECARB and MGSC projects.[75] The Development Phase large-scale injection projects are arguably akin to the large-scale carbon capture demonstration projects discussed above. They are needed to understand what actually happens to CO_2 underground when commercial-scale volumes are injected in the same or similar geologic reservoirs as would be used if CCS were deployed nationally.

In addition to understanding the technical challenges to storing CO_2 underground without leakage over hundreds of years, DOE also expects that the Development Phase projects will provide a better understanding of regulatory, liability, and ownership issues associated with commercial-scale CCS.[76] These nontechnical issues are not trivial, and could pose serious challenges to widespread deployment of CCS even if the technical challenges of injecting CO_2 safely and in perpetuity are resolved. For example, a complete regulatory framework for managing the underground injection of CO_2 has not been developed in the United States. However, EPA promulgated a rule under the authority of the Safe Drinking Water Act (SDWA) that creates a new class of injection wells under the existing Underground Injection Control Program. The new class of wells (Class VI) establishes national requirements specifically for injecting CO_2 and protecting underground sources of drinking water. EPA's stated purpose in proposing the rule was to ensure that CCS can occur in a safe and effective manner in order to enable commercial-scale CCS to move forward.[77]

The development of the regulation for Class VI wells highlighted that EPA's authority under the SDWA is limited to protecting underground sources of drinking water but does not address other major issues. Some of these include the long-term liability for injected CO_2, regulation of potential emissions to the atmosphere, legal issues if the CO_2 plume migrates underground across state boundaries, private property rights of owners of the surface lands above the injected CO_2 plume, and ownership of the subsurface reservoirs (also referred to as pore space).[78] Because of these issues and others, there are some indications that broad community acceptance of CCS may be a challenge. The large-scale injection tests may help identify the key factors that lead to community concerns over CCS, and help guide DOE, EPA, other agencies, and the private sector towards strategies leading to the widespread deployment of CCS. Currently, however, the general public is largely unfamiliar with the details of CCS and these challenges have yet to be resolved.[79]

FUTURE OUTLOOK

The success of the Clean Coal Program will ultimately be judged by the extent to which emerging technologies get deployed in domestic and international marketplaces. Both technical and financial challenges associated with the deployment of new "high risk" coal technologies must be overcome in order to be capable of achieving success in the marketplace. Commercial scale demonstrations help the industry understand and overcome startup issues, address component integration issues, and gain the early learning commercial experience necessary to reduce risk and secure private financing and investment for future plants.[80]

The testimony quoted above from Scott Klara of the National Energy Technology Laboratory sums up a crucial metric to the success of the federal CCS RD&D program, namely, whether CCS technologies are deployed in the commercial marketplace. To date, there are no commercial ventures in the United States that capture, transport, and inject large quantities of CO_2 (e.g., 1 million tons per year or more) solely for the purposes of carbon sequestration.

However, the CCS RD&D program in 2012 is just now embarking on commercial-scale demonstration projects for CO_2 capture, injection, and storage. The success of these demonstration projects will likely bear heavily on the future outlook for widespread deployment of CCS technologies as a strategy for preventing large quantities of CO_2 from reaching the atmosphere while plants continue to burn fossil fuels, mainly coal. Congress may wish to carefully review the results from these demonstration projects as they progress in order to gauge whether DOE is on track to meet its goal of allowing for an advanced CCS technology portfolio to be ready by 2020 for large-scale demonstration and deployment in the United States. In addition to the issues and programs discussed above, other factors might affect the demonstration and deployment of CCS in the United States. The use of hydraulic fracturing techniques to extract unconventional natural gas deposits recently has drawn national attention to the possible negative consequences of deep well injection of large volumes of fluids. Hydraulic fracturing involves the high-pressure injection of fluids into the target formation to fracture the rock and release natural gas or oil. The injected fluids, together with naturally occurring fluids in the shale, are referred to as produced water. Produced waters are pumped out of the well and disposed of. Often the produced waters are disposed of by re-injecting them at a different site in a different well. These practices have raised concerns about possible leakage as fluids are pumped into and out of the ground, and about deep well injection causing earthquakes. Public concerns over hydraulic fracturing and deep-well injection of produced waters may spill over into concerns about deep-well injection of CO_2. How successfully DOE is able to address these types of concerns as the large-scale demonstration projects move forward into their injection phases could affect the future of CCS deployment.

End Notes

[1] U.S. Department of Energy, National Energy Technology Laboratory, *Carbon Sequestration Program: Technology Program Plan*, Enhancing the Success of Carbon Capture and Storage Technologies, February 2011, p. 10, http://www.netl.doe.gov/technologies /carbon_seq/refshelf/2011_Sequestration_Program_Plan.pdf.

[2] U.S. Department of Energy, National Energy Technology Laboratory, *DOE/NETL Carbon Dioxide Capture and Storage RD&D Roadmap*, December 2010, http://www.netl.doe.gov /technologies/carbon_seq/refshelf/ CCSRoadmap.pdf. Hereinafter referred to as the DOE 2010 *CCS Roadmap*.

[3] EPA Fact Sheet: *Proposed Carbon Pollution Standard for New Power Plants*, http://epa.gov /carbonpollutionstandard/pdfs/20120327factsheet.pdf.

[4] Ibid. According to EPA, new power plants that use CCS would have the option to use a 30-year average of CO2 emissions to meet the standard, rather than meeting the annual standard each year. Under this option, new plants would be allowed to emit 1,800 pounds per megawatt-hour for the first 10 years of operation (a standard that should be achievable by an efficient supercritical coal-fired facility or an integrated gasification combined-cycle plant), provided that the facility committed to a 600 pound per megawatt-hour standard for the following 20 years of operation.

[5] DOE 2010 *CCS Roadmap*, p. 3.

[6] Andrew Childers, Bloomberg BNA, "EPA Carbon Rule for Utilities Expected to Continue Trend Toward Natural Gas Plants," March 28, 2012, http://www.bna.com/epa-carbon-rule-n12884908631/.

[7] U.S. Environmental Protection Agency, *Regulatory Impact Analysis for the Proposed Standard of Performance for Greenhouse Gas Emissions for New Stationary Sources*, Electric Utility Generating Units, March 2012, pp. ES3-ES4, http://epa.gov/carbonpollutionstandard/pdfs/20120327proposalRIA.pdf.

[8] The proposed rule is for new power plants, and exempts existing power plants as well as plants that make "modifications" as defined under EPA's New Source Performance Standards. See 40 C.F.R. Part 60.

[9] P.L. 109-58, Title IX, Subtitle F, §963; 42 U.S.C. 16293.

[10] P.L. 110-140, Title VII, Subtitles A and B.

[11] Among other provisions, the bill would also amend EISA to expand the number of large CCS demonstration projects from seven to ten.

[12] DOE 2010 *CCS Roadmap*, p. 2.

[13] The Coal Program contains CCS RD&D activities, and is within DOE's Office of Fossil Energy, Fossil Energy Research and Development, as listed in DOE detailed budget justifications for each fiscal year. See, for example, U.S. Department of Energy, FY2013 Congressional Budget Request, volume 3, *Fossil Energy Research and Development*, http://www.cfo.doe.gov/budget/13budget/Content/Volume3.pdf.

[14] U.S. Department of Energy, FY2013 Congressional Budget Request, volume 3, *Fossil Energy Research and Development*, p. 411.

[15] DOE 2010 *CCS Roadmap*, p. 3.

[16] DOE states that the cost of deploying currently available CCS post-combustion technology on a supercritical pulverized coal-fired power plant would increase the cost of electricity by 80%. DOE 2010 *CCS Roadmap*, p. 3.

[17] DOE 2010 *CCS Roadmap*, p. 3.

[18] DOE 2010 *CCS Roadmap*, p. 2.

[19] DOE 2010 *CCS Roadmap*, p. 11.

[20] DOE 2010 *CCS Roadmap*, pp. 11-12.

[21] DOE 2010 *CCS Roadmap*, p. 12.

[22] See, for example, the presentations given by Edward Rubin of Carnegie Mellon University, Howard Herzog of the Massachusetts Institute of Technology, and Jeff Phillips of the Electric Power Research Institute, at the CRS seminar *Capturing Carbon for Climate Control: What's in the Toolbox and What's Missing*, November 18, 2009. (Presentations available from Peter Folger at 7-1517.) Rubin stated that at least 10 full-scale demonstration projects would be needed to establish the reliability and true cost of CCS in power plant applications. Herzog also called for at least 10 demonstration plants worldwide that capture and sequester a million metric tons of CO2 per year. In his presentation, Phillips stated that large-scale demonstrations are critical to building confidence among power plant owners.

[23] For example, an MIT report estimated that the costs of capture could be 80% or more of the total CCS costs. John Deutsch et al., *The Future of Coal*, Massachusetts Institute of Technology, An Interdisciplinary MIT Study, 2007, Executive Summary, p. xi.

[24] *The Future of Coal*, p. 97.

[25] For a fuller discussion of the relationship between costs of developing technologies analogous to CCS, such as SO2 and NOx scrubbers, see CRS Report R41325, *Carbon Capture: A Technology Assessment*, by Peter Folger.

[26] U.S. Department of Energy, *Strategic Plan*, May 2011, p. 18, http://energy.gov/sites/prod/files/ 2011_DOE_Strategic_Plan_.pdf.

[27] DOE had solicited and awarded funding for CCPI projects in two previous rounds of funding: CCPI Round 1 and Round 2. The Recovery Act funds were to be allocated CCPI Round 3,

focusing on projects that utilize CCS technology and/or the beneficial reuse of CO2. For more details, see http://www.fossil.energy.gov/programs/ powersystems/cleancoal/.

[28] U.S. Department of Energy, *Detailed Budget Justifications FY2010*, volume 7, Fossil Energy Research and Development, p. 35, http://www.cfo.doe.gov/budget/ 10budget/ Content/Volumes/Volume7.pdf.

[29] DOE 2010 *CCS Roadmap*, p. 15.

[30] The first solicitation closing date was January 20, 2009; the second solicitation closing date was August 24, 2009. Thus the first set of project proposals were submitted prior to enactment of the Recovery Act. See http://www.fossil.energy.gov/programs /powersystems/ cleancoal/.

[31] MIT Carbon Capture & Sequestration Technologies, *Plant Barry Fact Sheet: Carbon Dioxide Capture and Storage Project*, http://sequestration.mit.edu/tools/projects/plant_barry.html.

[32] Ibid.

[33] U.S. DOE, Fossil Energy Techline, *Secretary Chu Announces Two New Projects to Reduce Emissions from Coal Plants*, July 1, 2009, http://www.fossil.energy.gov/news/ techlines/ 2009/09043-DOE_Announces_CCPI_Projects.html.

[34] Lauren Donovan, "Basin Shelves Lignite's First Carbon Capture Project," *Bismarck Tribune*, December 17, 2010, http://bismarcktribune.com/news/local/a5fb7ed8-0a1b-11e0-b0ea-001cc4c03286.html.

[35] U.S. DOE, Fossil Energy Techline, *Secretary Chu Announces $3 Billion Investment for Carbon Capture and Sequestration*, December 4, 2009, http://www.fossil.energy Secretary_Chu_Announces_CCS_Invest.html.

[36] Matthew L. Wald and John M. Broder, "Utility Shelves Ambitious Plan to Limit Carbon," *New York Times*, July 13, 2011, http://www.nytimes.com/2011/07/14/business/energy-environment/utility-shelves-plan-to-capture-carbondioxide.html?_r=1.

[37] Michael G. Morris, chairman of AEP, quoted in *New York Times* article by Wald and Broder, July 13, 2011.

[38] Wald and Broder, *New York Times*, July 13, 2011.

[39] MIT Carbon Capture & Sequestration Technologies, *AEP Mountaineer Fact Sheet: Carbon Dioxide Capture and Storage Project*, http://sequestration.mit.edu/tools/projects/aep_ alstom_mountaineer.html.

[40] Telephone conversation with Joseph Giove, DOE Office of Fossil Energy, March 19, 2012.

[41] U.S. DOE Fossil Energy Techline, "Secretary Chu Announces Up To $154 Million for NRG Energy's Carbon Capture and Storage Project in Texas," March 9, 2010, http://www.fossil. energy.gov/news/techlines/2010/10005- NRG_Energy_Selected_to_Receive_DOE.html.

[42] Telephone conversation with Joseph Giove, DOE Office of Fossil Energy, April 11, 2011.

[43] U.S. Congress, House Committee on Appropriations, Subcommittee on Military Construction, Veterans Affairs, and Related Agencies, *Military Construction and Veterans Affairs and Related Agencies Appropriations Act, 2012*, conference report to accompany H.R. 2055, 112th Cong., 1st sess., December 15, 2011, H.Rept. 112-331 (Washington: GPO, 2011), p. 851.

[44] Telephone conversation with Joseph Giove, DOE Office of Fossil Energy, March 19, 2012.

[45] Email from Regis K. Conrad, Director, Division of Cross-Cutting Research, DOE, March 20, 2012.

[46] U.S. DOE, National Energy Technology Laboratory, *Major Demonstrations, Industrial Capture and Storage (ICCS): Area 1*, http://www.netl.doe.gov/technologies/coalpower/ cctc/iccs1/index.html#.

[47] U.S. DOE, *Recovery Act, Innovative Concepts for Beneficial Reuse of Carbon Dioxide*, http://fossil.energy.gov/ recovery/projects/beneficial_reuse.html.

[48] Email from Regis K. Conrad, Director, Division of Cross-Cutting Research, DOE, March 20, 2012.

[49] U.S. DOE, Carbon Capture and Storage from Industrial Sources, Industrial Carbon Capture Project Selections, http://fossil.energy.gov/recovery/projects/iccs_projects_0907101.pdf.

[50] See http://www.fossil.energy.gov/news/techlines/2008/08003-DOE_Announces_ Restructured_FutureG.html.

[51] Prior to when DOE first announced it would restructure the program in 2008, the FutureGen Alliance announced on December 18, 2007, that it had selected Mattoon, IL, as the host site from a set of four finalists. The four were Mattoon, IL; Tuscola, IL; Heart of Brazos (near Jewett, TX); and Odessa, TX.

[52] See DOE announcement on June 12, 2009, http://www.fossil.energy.gov/ news/techlines /2009/09037- DOE_Announces_FutureGen_Agreement.html.

[53] See DOE Techline, http://www.netl.doe.gov/publications/press/2010/10033- Secretary_Chu_ Announces_FutureGen_.html.

[54] Ameren had pla nned to replace the oil-fired boiler with a coal-fired boiler using oxy-combustion technology to allow carbon capture. See http://www.futuregenalliance.org/pdf/ FutureGen%20FAQ-General%20042711.pdf.

[55] Environmental News Service, "Future Unclear for FutureGen 2.0 Carbon Capture and Storage Network," August 17, 2010, http://www.ens-newswire.com/ens/aug2010/2010-08-17-093.html. Also, see David Mercer, "Ill. Town Rejects Changes to Federal FutureGen Coal Project, Backs Out After More Than Two Years," *Associated Press*, August 11, 2010, http://www.startribune.com/templates/Print_This_Story?sid=100479394.

[56] DOE Techline, *FutureGen Industrial Alliance Announces Carbon Storage Site Selection Process for FutureGen 2.0*, http://www.fossil.energy.gov/news/techlines/2010/10051-FutureGen_Site_Selection_Process_A.html.

[57] FutureGen Alliance, "FutureGen Alliance Selects Morgan County, Ill. as the Site for the FutureGen 2.0 Carbon Storage Facility," February 28, 2011, http://www.futuregena lliance.org/pdf/pr_02_28_11.pdf. In its announcement, the FutureGen Alliance notes that it selected Christian County and Douglas County as alternative sites in case of concerns over the Morgan County selection.

[58] NETL factsheet, *FutureGen 2.0*, http://www.netl.doe.gov/publications/factsheets/ project/ FE0001882- FE0005054.pdf.

[59] Ibid.

[60] Ibid.

[61] "Ameren Will Shutter 2 Ill. Plants," *Greenwire*, October 4, 2011, http://www.eenews. net/Greenwire/2011/10/04/ archive/12?terms=futuregen.

[62] Matthew Wald, "Coal Project Hits Snag as a Partner Backs Off," *New York Times*, November 10, 2011,http://www.nytimes.com/2011/11/11/business/energy-environment/coal-project-hits-snag-as-a-partner-backs-off.html? r=1.

[63] Gabriel Nelson, "FutureGen in Talks to Buy Ill. Power Plant for CCS Pilot Project," *Greenwire*, November 28, 2011, http://www.eenews.net/Greenwire/2011/11/28/archive/2? terms=futuregen.

[64] See, for example, the experience in Germany by Swedish power utility Vattenfall, which recently cancelled plans to build a CCS power plant because of local opposition to injecting and storing CO2 underground. Paul Voosen, "Public Outcry Scuttles German

Demonstration Plant," *Greenwire*, December 6, 2011, http://www.eenews.net/Greenwire/ 2011/12/06/archive/10?terms=futuregen.

[65] E. S. Rubin, "The Government Role in Technology Innovation: Lessons for the Climate Change Policy Agenda," Institute of Transportation Studies, 10th Biennial Conference on Transportation Energy and Environmental Policy, University of California, Davis, CA (August 2005).

[66] National Aeronautics and Space Administration, "Definition of Technology Readiness Levels," at http://esto.nasa.gov/files/TRL_definitions.pdf.

[67] For a more thorough discussion of different schemes describing stages of technology development, see chapter 4 of CRS Report R41325, *Carbon Capture: A Technology Assessment*, by Peter Folger.

[68] E. S. Rubin, "The Government Role in Technology Innovation: Lessons for the Climate Change Policy Agenda," Institute of Transportation Studies, 10th Biennial Conference on Transportation Energy and Environmental Policy, University of California, Davis, CA (August 2005).

[69] Another possible source of uncertainty for FutureGen, and other large industrial CCS projects, is cost recovery during the operating phase of the plant after the construction phase and initial capital investments are made. "Learning by doing" should increase operating efficiency, but it is unclear by how much and over what time span. For more discussion on cost trajectories and expected efficiency gains, see CRS Report R41325, *Carbon Capture: A Technology Assessment*, by Peter Folger.

[70] The total DOE share for the 10 projects is $46.6 million. See DOE, *Recovery Act*, http://fossil.energy.gov/recovery/ projects/site_characterization.html.

[71] DOE 2010 *CCS Roadmap*, p. 55.

[72] Four Canadian provinces are partners with DOE in two of the regional partnerships, and are members with other participating organizations that are contributing funding and other support to the partnerships.

[73] DOE National Energy Technology Laboratory, Carbon Sequestration Regional Carbon Sequestration Partnerships, http://www.netl.doe.gov/technologies/ carbon_seq/infra structure/rcsp.html.

[74] Ibid.

[75] For details on the two large-scale injection experiments by SECARB, see http://www.secarbon.org/; for details on the large-scale injection experiment by MGSC, see http://sequestration.org/.

[76] DOE National Energy Technology Laboratory, *Carbon Sequestration Regional Partnership Development Phase (Phase III) Projects*, http://www.netl.doe.gov/technologies/ carbon_seq/infrastructure/rcspiii.html.

[77] For more information on the EPA Class VI wells in particular, and the Safe Drinking Water Act generally, see CRS Report RL34201, *Safe Drinking Water Act (SDWA): Selected Regulatory and Legislative Issues*, by Mary Tiemann.

[78] For a discussion of several of these legal issues, see CRS Report RL34307, *Legal Issues Associated with the Development of Carbon Dioxide Sequestration Technology*, by Adam Vann and Paul W. Parfomak.

[79] For more information on the different issues regarding community acceptance of CCS, see CRS Report RL34601, *Community Acceptance of Carbon Capture and Sequestration Infrastructure: Siting Challenges*, by Paul W. Parfomak.

[80] Testimony of Scott Klara, Deputy Laboratory Director, National Energy Technology Laboratory, U.S. Department of Energy, in U.S. Congress, Senate Energy and Natural

Resources Committee, *Carbon Capture and Sequestration Legislation*, hearing to receive testimony on carbon capture and sequestration legislation, including S. 699 and S. 757, 112th Cong., 1st sess., May 12, 2011, S.Hrg. 112-22.

In: Carbon Capture and Sequestration   ISBN: 978-1-62257-810-8
Editors: A.C. Mitchell and R. Freeman   © 2013 Nova Science Publishers, Inc.

Chapter 3

FEDERAL EFFORTS TO REDUCE THE COST OF CAPTURING AND STORING CARBON DIOXIDE*

Congressional Budget Office

NOTES

Numbers in the text and tables may not add up to totals because of rounding.

The diagram on the cover is reproduced from Michael Mann and Lee R. Kump, *Dire Predictions: Understanding Global Warming*, 1st ed., © 2009. Reprinted by permission of Pearson Education, Inc., Upper Saddle River, New Jersey.

SUMMARY

Electricity generation in the United States depends heavily on the use of coal: Coal-fired power plants produce 40 percent to 45 percent of the nation's electricity. At the same time, those facilities account for roughly a third of all U.S. emissions of carbon dioxide (CO_2), which together with

---

* This is an edited, reformatted and augmented version of a Congressional Budget Office No. 4146 publication, dated June 2012.

other greenhouse gases has become increasingly concentrated in the atmosphere. Most climate scientists believe that the buildup of those gases could have costly consequences.

One much-discussed option for reducing the nation's greenhouse gas emissions while preserving its ability to produce electricity at coal-fired power plants is to capture the CO_2 that is emitted when the coal is burned, compress it into a fluid, and then store it deep underground. That process is commonly called carbon capture and storage (CCS). Although the process is in use in some industries, no CCS-equipped coal-fired power plants have been built on a commercial scale because any electricity generated by such plants would be much more expensive than electricity produced by conventional coal-burning plants. Utilities, rather than federal agencies, make most of the decisions about investments in the electricity industry, and today they have little incentive to equip their facilities with CCS technology to lessen their CO_2 emissions.

Since 2005, lawmakers have provided the Department of Energy (DOE) with about $6.9 billion to further develop CCS technology, demonstrate its commercial feasibility, and reduce the cost of electricity generated by CCS-equipped plants. But unless DOE's funding is substantially increased or other policies are adopted to encourage utilities to invest in CCS, federal support is likely to play only a minor role in deployment of the technology.

Engineers have estimated that, on average, electricity generated by the first CCS-equipped commercial-scale plants would initially be about 75 percent more costly than electricity generated by conventional coal-fired plants. (Most of that additional cost is attributable to the extra facilities and energy that would be needed to capture the CO_2.) That initial cost differential would probably shrink, however, as the technology became more widely applied and equipment manufacturers and construction companies became more familiar with it— a pattern of cost reduction called learning-by-doing.

DOE aims to bring down the additional costs for generating electricity with CCS technology to no more than 35 percent, or less than half the current cost premium. Such a cost differential, if combined with a tax on carbon or policies restricting CO_2 emissions, could allow coal-fired plants with CCS to be competitive with those without CCS.

Such a reduction in costs might be accomplished over time through learning-by-doing, which would require that a certain amount of new generating capacity be built—in the form of new coal-fired CCS-equipped generating plants. Using the historical pace of reductions in costs for earlier emission-control technologies, the Congressional Budget Office (CBO) estimates that more than 200 gigawatts (GW) of coal-fired generating capacity with CCS capabilities will have to be built to meet DOE's cost reduction goal. That estimate of new capacity, which is equivalent to about two-thirds of the total current capacity of U.S. coal-powered electricity generation plants, is subject to considerable uncertainty. Nevertheless, in the absence of a significant technological breakthrough, it seems clear that a large amount of new CCS capacity—installed either at new plants or, through retrofitting, at existing plants—would be needed to reduce costs by enough to achieve DOE's goal.

But the demand for electricity in the United States is growing slowly, and even if DOE's cost reduction target was attained, coal-fired power plants equipped with CCS technology would not be competitive with coal-fired plants that lacked it unless policies restricting CO2 emissions or imposing a price on them were adopted. Consequently, under current laws and policies, utilities are unlikely to build that much new generating capacity—that is, more than 200 GW—or invest in adding CCS technology to much of their existing capacity for many decades. If, however, new policies restricted or imposed a price on CO_2 emissions, the domestic stock of electricity generation plants would turn over more rapidly, and CCS technology would become more competitive economically, increasing the potential for construction of CCS-equipped plants in the United States. Nevertheless, investors already have several options for generating electricity—nuclear power, wind, biomass, other renewables, and natural gas—that produce few, if any, CO_2 emissions. The amount of investment in CCS would depend on how costs for the different alternatives compared with costs for electricity generation without CCS.

Reductions in costs for CCS-equipped power plants could also come from experience outside the United States. Demand for electricity is growing rapidly in other parts of the world—for example, China and India—and those countries are increasing their capacity to satisfy it. If plants that were equipped with early versions of the CCS technology were built abroad or if some coal-fired power plants now in operation in other countries were retrofitted with CCS, the cost of generating electricity at plants that were subsequently built or retrofitted in the United States would be expected to be lower than the cost of generating electricity at the plants that were built initially. At present, however, foreign investment in CCS, like investment in the technology in the United States, centers not on building full-scale CCS-equipped commercial plants but on conducting research and development, carrying out small-scale demonstrations of the technology's feasibility, and building pilot plants.

Until now, most efforts to develop CCS have focused on coal-fired power plants. However, the price of natural gas has dropped substantially in recent years, and the share of electricity generated by natural gas-fired plants has expanded and is likely to continue to grow. The cost of producing electricity with a natural gas-fired plant equipped with CCS could be lower, depending on future prices for coal and natural gas, than the cost of producing electricity with a coal-fired CCS plant. At present, though, regulatory action to curb CO_2 emissions is more likely to shift electricity production from coal to natural gas (without CCS) and other low-emission fuels, such as biomass, rather than to CCS-capable plants.

CBO's analysis suggests that unless the federal government adopts policies that encourage or require utilities to generate electricity with fewer greenhouse gas emissions, the projected high cost of using CCS technology means that DOE's current program for developing CCS is unlikely to do much to support widespread use of the technology. A number of other policy approaches could be considered. For example, lawmakers could redirect resources that now fund technology demonstration projects toward research and development, for which the rationale for federal involvement is strongest and the record of success better. Alternatively, policymakers could impose costs on users of electricity whose generation releases greenhouse gases—for example, through a tax on carbon—thereby making CCS more competitive, or they could experiment with different types of electricity production subsidies that would provide more incentive for private-sector investments in CCS. As another option, lawmakers could reduce or eliminate future spending for CCS, leaving most of the potential for further development of CCS technology to countries with high rates of growth in the demand for electricity and in the need for new electricity-generating capacity.

INTRODUCITON

Concerns about global warming have raised questions about the United States' continued dependence on coal for producing electricity. About 1,400 coal-fired generating units located in roughly 600 power plants produce 40 percent to 45 percent of the electricity generated annually in this country and in so doing release about a third of the carbon dioxide attributable to human activities in the United States each year.[1] The consensus among scientific experts is that increasing concentrations of greenhouse gases in the atmosphere—including CO_2, which is the most common—are likely to have extensive, highly uncertain but potentially costly effects on regional climates throughout the world.[2]

The federal government, through the Department of Energy, is seeking ways to reduce greenhouse gas emissions while preserving the nation's ability to continue to rely on coal to produce electricity. A policy to reduce CO_2 emissions would benefit the United States by lessening the risk of costly changes to the climate. However, such a policy would also impose costs on the U.S. economy because it would limit activities that produce those emissions.

Depending on the type of policy that lawmakers chose, electric utilities and their customers, coal producers, or certain areas of the country could bear increased costs or a considerable loss of income and jobs.[3] As a result, policymakers have sought options that would reduce CO_2 emissions but also limit the potential impact on the economy and allow the nation to continue to produce electricity from coal. Since 2005, lawmakers have provided DOE with about $6.9 billion to develop and demonstrate the commercial feasibility of technologies that would allow coal-burning power plants to generate electricity without emitting CO_2 into the atmosphere. Instead, the CO_2 would be removed from a plant's exhaust stream, compressed into a liquid, and stored underground indefinitely. Collectively, those processes are usually called carbon capture and storage.

This Congressional Budget Office study examines current federal policies that support the development, demonstration, and deployment of CCS technology and the policies' potential to reduce the future costs of generating electricity with power plants that capture and store carbon.

CARBON CAPTURE
AND STORAGE TECHNOLOGY

Any industrial process that produces CO_2 can be modified to capture and store it. For example, CCS technology can be applied to coal-fired power plants—the primary focus of this report—as well as to generating facilities fueled by natural gas; it can also be used in manufacturing such products as cement, ethanol, and fertilizer. In coal-fired power plants, CCS requires facilities and processes that accomplish the following tasks:

- Capture CO_2 at the plant and compress it into a liquid;
- Transport the compressed CO_2, usually through a pipeline, to a storage site; and
- Store CO_2 deep underground in a porous rock formation (see Figure 1).

The feasibility of using such facilities and processes in the generation of electricity has been explored on a small scale, but the technology has yet to be widely adopted.

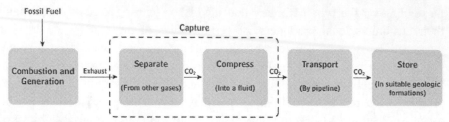

Source: Congressional Budget Office based on CCSReg Project, Carbon Capture and Sequestration: Framing the Issues for Regulation (January 2009, updated March 2009), p. 10.

Figure 1. Steps in the Capture and Storage of Carbon Dioxide After Electricity Generation at a Coal-Fired Power Plant.

Capturing Carbon Dioxide

When CCS technology is used in coal-fired power plants, exhaust gases that contain CO_2 are streamed through (or otherwise put into contact with) a specialized material that absorbs most of the CO_2 while allowing the rest of the exhaust to pass. Subsequently, the CO_2 that was absorbed is extracted, by heating or otherwise treating that material, and then compressed into a liquid. Because the approach applies CCS technology to exhaust gases, it is termed a postcombustion method. Two alternative approaches have been proposed and in some instances used in pilot plants or in fields unrelated to electricity generation—for example, in making glass. One alternative approach is to capture CO_2 through a precombustion method such as gasification. That process separates the CO_2 by transforming the coal into a gas (often referred to as syngas) before it is burned. The other alternative approach is oxy-fuel combustion, in which coal is burned in pure oxygen rather than air to produce exhaust gas that consists primarily of water and CO_2 and that is ready for drying, compression, and storage. However, CBO's analysis focuses on the postcombustion approach because that technology is the only one that is compatible with the most commonly used designs for electricity-generating plants.

The process of treating the absorbent material and compressing the CO_2 consumes a great deal of energy—so much so that the capture and compression of CO_2 reduce the net amount of energy that the power plant yields for customers by between 15 percent and 30 percent. Thus, a plant equipped with CCS technology must be larger than a traditionally equipped plant and must burn more coal to serve the same number of customers. Engineering studies suggest that the capture portion of the process will account for approximately 90 percent of the additional costs required to construct and operate a plant that uses CCS instead of conventional technology.

Transporting Carbon Dioxide

Once compressed, the captured CO_2 must be transported to an underground storage site. If CCS was widely adopted, it would be necessary to substantially expand the existing pipeline network to transport the gas to storage sites that might be hundreds of miles away. Such a network could use pipeline technology that has already been developed to transport carbon dioxide to oil fields, where it is injected into wells to boost their production— a process known as enhanced oil recovery. Currently, about 4,000 miles of pipeline is used for that purpose.

The market in CO_2 for enhanced oil recovery and the network of pipelines to transport compressed gas are expanding, a trend reflected in forecasts by DOE's Energy Information Administration (EIA) that enhanced oil recovery will increase significantly over the next 25 years.[4]

Storing Carbon Dioxide

After it has been transported, the CO_2 captured at power plants would be injected through deep wells, similar to oil or natural gas wells, into porous geologic formations roughly a kilometer or more underground. (That depth is considered to be enough to maintain the pressure required to keep CO_2 in its liquid state.) The types of underground formations typically discussed for storing CO_2 include depleted oil and natural gas reservoirs, porous formations filled with brine, and unminable coal seams. The technology for underground storage of CO_2 is well developed. Enhanced oil recovery efforts, together with some other storage projects, have already pumped more than 500 million tons of CO_2 underground, most of which has remained in place.[5]

Geologists believe that geologic formations in the United States have the potential to store enough CO_2 to permit widespread use of CCS technology in the country's electric power industry. Current estimates by DOE and the International Energy Agency suggest a theoretical storage potential of over

3,000 billion metric tons, or roughly 1,000 years' worth of the CO_2 emitted by U.S. coal-fired utilities.[6] The U.S. Geological Survey and DOE are trying to refine those estimates to determine how much of that potential could be developed.

Status of CCS Technology Development

The technology for separating and capturing CO_2 is already in use in several industries, although the conditions in which it is employed are less demanding than those that apply in utility-scale coal-fired power plants. CCS is used, for example, by producers of natural gas, which, as it comes out of the ground, routinely contains more CO_2 than conventional uses allow. To make natural gas fit for sale, refiners separate the CO_2 from the gas and simply vent it into the atmosphere. More recently, facilities that produce natural gas have shown that large-scale capture of CO_2 is technically and operationally feasible and that the captured gas can be compressed into a liquid, transported, and stored underground. The Sleipner natural gas plant in the North Sea off the coast of Norway and the Salah Gas plant in Algeria are the most prominent examples of that use of CCS technology. Both of those plants strip the CO_2 from natural gas and pump it into suitable geologic formations. Similarly, the Great Plains Synfuels Plant uses a precombustion CCS technology that turns coal into a gas. The plant, located near Beulah, North Dakota, is the largest CCS-equipped plant in the world; its early version of the technology has a CO_2 capture rate of 50 percent. (One of DOE's CCS-related goals is to have CCS-equipped plants capture 90 percent of the CO_2 emitted during the production of electricity.) The resulting liquid is piped to Canada, where it is used in enhanced oil recovery.

Technology for capturing CO_2 has been used in other industries, although not on a large scale. Some producers of ammonia, hydrogen, and ethanol may separate CO_2 from other gases as part of their production process. Also, in the electricity generation industry, a few facilities separate and capture a fraction of their CO_2 emissions for sale to nearby food-processing plants—to be used, for example, in carbonated products. Those facilities are relatively small, however, and the CCS technology they use is not of a sufficient scale to eliminate all or most of the CO_2 emissions from the exhaust of a commercial power plant.

Because CO_2 capture technologies are already in use to some extent, industrial engineers may have already taken advantage of the available opportunities to make the capture process more efficient and hence less costly. The additional advances necessary to markedly improve the technology's performance or reduce its costs are likely to require substantial investment in research and development (R&D) as well as experience in building and operating new plants. The area of research that is widely thought by experts to have the greatest potential to reduce costs and so move the technology toward more widespread use is to improve the absorbent materials used to extract the CO_2 and thereby reduce the energy required for the capture stage of the process.

Table 1. Large-Scale Projects to Install CCS Technology in Power Plants That Are Currently Planned or Under Construction in the United States

Project	Private-Sector Project Leader	Location	Size (Megawatts)	DOE Funding (Millions of dollars)	Planned Completion Date
FutureGen 2.0	FutureGen Industrial Alliance	Meredosia, Ill.	200	1,000	2015
Hydrogen Energy California	SCS Energy	Kern County, Calif.	390	308	2014
Kemper County	Mississippi Power/Southern Company	Kemper County, Miss.	582	270	2014
Tenaska Trailblazer Energy Center	Tenaska, Inc.	Sweetwater, Tex.	600	0	2014
Texas Clean Energy	Summit Power Group	Ector County, Tex.	400	450	2014–2015
W.A. Parish Plant	NRG Energy	Thompsons, Tex.	60	154	2017

Source: Congressional Budget Office based on Carbon Capture and Sequestration Technologies at MIT, CCS Project Database, "Power Plant Carbon Dioxide Capture and Storage Projects: Large-Scale Power Plant CCS Projects Worldwide" (March 15, 2012), http://sequestration.mit.edu/tools/projects/index_capture.html.

Note: CCS = carbon capture and storage (the set of processes and technologies that separate carbon dioxide, or CO2, from other gases generated when a fossil fuel is burned; compress the CO2 into a fluid; and then transport it to an underground location for storage); DOE = Department of Energy.

However, integrating CCS technology into the production of electricity—and specifically into electricity generation at coal-fired power plants—appears to be more demanding technically than, for example, the use of CCS in the production of natural gas. That added technical complexity, which contributes to the greater cost of electricity generation at CCS-equipped versus conventional power plants, has limited the technology's use in existing coal-fired facilities.

Until relatively recently, though, power companies worldwide were planning to undertake roughly 30 new large-scale coal-powered CCS projects—10 of them in the United States—to demonstrate the technology's commercial viability.

The planned facilities ranged in capacity from 60 megawatts (MW) to more than 680 MW.[7] (The average coal-fired generating unit in the United States has a summer capacity of about 230 MW.)[8]

Plans for many of those "demonstration" projects have now changed, however, and in fact, a number of them (six in the European Union, Canada, and Norway, and four in the United States) have already been canceled or put on hold.

Several projects in the United States were started when adoption of a cap-and-trade program or some other U.S. policy for reducing greenhouse gases seemed more likely to investors than it does today. As the prospect of a nationwide emission-reduction program has faded, so have plans for new CCS-equipped power plants, and now only six large-scale demonstration projects are still planned or under construction in the United States (see Table 1).

Of those six projects, five have received federal funding. The private-sector organizers of the remaining project have discussed their hope of receiving federal funds, but as of this writing, the project has not received any federal support.

Table 2. Funding for the Department of Energy's Coal Programs
(Billions of dollars of budget authority)

	2005	2006	2007	2008	2009	2010	2011	2012
Coal Program Funding	0.3	0.4	0.4	0.5	4.1	0.4	0.4	0.4

Source: Congressional Budget Office based on information from Department of Energy, Office of Fossil Energy, "Our Budget" (various years), www.fossil.energy.gov/aboutus/budget/index.html; and U.S. House of Representatives, Conference Report to Accompany H.R. 1, House Report 111-16 (February 12, 2009), p. 428.

Note: Budget authority is the authority provided by law to incur financial obligations that will result in immediate or future outlays of funds by the federal government.

Federal Policy to Demonstrate Current CCS Technology and Promote Its Future Technological Development

DOE encourages the advancement of CCS technologies through spending for projects to demonstrate the feasibility of using current CCS technology in commercial-scale power plants and by funding R&D activities to develop future generations of more advanced and more efficient equipment. In addition, the federal government encourages the use of CCS by providing tax credits for private utilities that choose to invest in and produce electricity from CCS-equipped plants.

The Department of Energy's CCS Programs

DOE's programs devoted to CCS have two overall goals: to reduce the added costs for electricity produced by CCS-equipped coal-fired power plants to no more than 35 percent and to improve the technology so that CCS-equipped plants capture 90 percent of the CO_2 emitted during the electricity generation process.[9]

If the costs for CCS could be reduced to that extent, policies that imposed costs on emissions of carbon, such as a tax, might then make CCS-equipped plants competitive with conventional coal-fired facilities. Achieving those goals would allow for substantial reductions in CO_2 emissions while still permitting the extensive use of coal in the generation of power. The desired amount of cost reduction, DOE argues, would require "aggressive but feasible" development of the technology and would, in the end, reduce the cost of meeting targets for CO_2 emissions by billions of dollars.

The programs to develop and promote CCS technology —the CCS Demonstrations Program and the Carbon Capture and Storage and Power Systems Program—are overseen by DOE's Office of Fossil Energy. That office received about $0.4 billion for its coal programs in 2012 and similar amounts in most of the previous several years (see Table 2). It also received a large infusion of funds from the American Recovery and Reinvestment Act of 2009, which provided $3.4 billion for DOE's CCS efforts (bringing total funding in 2009 for that purpose to $4.1 billion). Much of the money appropriated for CCS remains unspent, however, in part because of the normal time lags in designing and building large projects but also because private investors have canceled several projects for which the federal government was planning to provide some funding.*Full-Scale Demonstration Projects.* DOE is participating in five of the six full-scale CCS demonstration projects currently being planned and built in the United States (see Table 1).

As of April 2012, DOE had committed $2.2 billion to the construction of those plants; private parties are contributing roughly $10 billion, although that amount does not take into account tax advantages and other considerations from state and local governments. The plants that DOE is helping fund generally embody new technology not only to capture and store CO_2 but also to advance the efficiency of coal-fired electricity production more broadly. Most of those projects include plans to use the captured CO_2 for enhanced oil recovery as a way to increase the projects' economic returns.

One DOE demonstration project that has evoked a great deal of public comment is the FutureGen project, to be located in Illinois. The FutureGen Industry Alliance, composed of several power companies, leads the project; its scope includes retrofitting an existing oil-fueled boiler with a coal-fueled boiler equipped with oxy-fuel combustion technology (for burning coal in pure oxygen) that could generate 200 MW of power. (The plans for the FutureGen project have been substantially modified since the program was first proposed in 2003, and the current project is now designated FutureGen 2.0.) In addition, the project calls for construction of a pipeline and storage facilities sufficient to accommodate an estimated 1.3 million tons per year of compressed carbon dioxide.

According to the FutureGen Industry Alliance, the project will cost approximately $1.3 billion, most of which will be covered by a $1 billion grant provided by DOE.[10]

The alliance plans to spend approximately $730 million to retrofit the boiler and $550 million to build the pipeline and storage facilities. Construction is slated to begin in late 2012, and completion of the project is planned for 2015.

In addition to its support for the five plants currently under way, DOE had committed an additional $730 million to three of the four large-scale CCS projects that have been postponed or canceled. Much of that funding was intended for the Mountaineer power plant in New Haven, West Virginia. The large utility American Electric Power recently finished a CCS pilot project there and had planned to expand its CCS capabilities to encompass 20 percent of the plant's electricity generation. DOE was to have provided a grant of $335 million to pay for half the cost of the expansion. But the company has postponed construction indefinitely, citing uncertainty about federal policies for reducing greenhouse gases and prohibitions by local utility commissions on increases in rates (which would have prevented the utility from recouping its costs for the plant's construction).

Research and Development Programs. DOE's CCSrelated R&D activities have focused mainly on capturing and storing CO_2. The department's analysts believe that the current technology for capturing CO_2 could never meet DOE's goal of reducing the cost of CCS-generated electricity. Consequently, DOE has been seeking to develop next-generation CCS equipment and processes that would capture CO_2 more quickly and more completely but use less energy than today's technology does. For example, some DOE-sponsored research involves basic and applied studies to identify better materials for absorbing CO_2 and reducing the amount of energy used by the process for capturing the gas.

Those projects to develop new technologies are expected to begin to reach the pilot and demonstration stages over the next 10 to 15 years. DOE is also sponsoring research on reducing the cost and increasing the reliability and efficiency of CCS-equipped coal-gasification plants.

In the area of storing carbon dioxide, DOE is funding research to develop techniques to enhance firms' ability to predict the movement of CO_2 underground. Such techniques include the use of software that would enable analysts to better understand the capacity of underground formations for storing the compressed gas.

Federal Tax Preferences

Federal support for CCS extends beyond DOE's coal programs to provisions of law that reduce the amount of taxes paid by utilities that invest in the technology and use it to generate electricity.

The Energy Policy Act of 2005 and the Energy Improvement and Extension Act of 2008 (Division B of Public Law 110-343) created several different tax credits for investment in plants that incorporate various types of "clean coal" technology.

For operators of a plant to be eligible for those credits, the Secretary of the Treasury, in consultation with the Secretary of Energy, must certify that the plant will capture and store at least 65 percent to 75 percent of its total CO_2 emissions, depending on the specific technology and credit. Altogether, lawmakers authorized almost $3 billion in investment tax credits, which, the Joint Committee on Taxation forecasts, will cost roughly $0.2 billion per year in forgone revenues through 2015.[11]

Given the lack of new or proposed projects and the cancellation of projects already under way, a substantial portion of the authorized credits will probably never be used.

The Cost of Producing Electricity Using CCS Technology

CBO has compared the estimated costs of producing electricity at conventional coal-fired power plants with those that would be incurred at facilities equipped with CCS technology. Initially, the cost of generating electricity at a new coal-fired CCS-equipped plant would be substantially higher than the cost of generating it at a plant that produced the same net output of electricity but used conventional technology to do it. However, that premium could decline over time as electric utilities gained experience in installing and using CCS, a pattern seen with other new technologies. Even so, reducing the cost by enough to achieve DOE's goal of only a 35 percent premium could require a lengthy process of building a large amount of new electricity generation capacity.

Cost Differentials Associated with Current CCS Technology

Analysts have assessed what happens to the cost of generating electricity when CCS equipment is added to a power plant. However, because no full-sized CCSequipped plants have been built, all of the estimates that analysts have produced derive from engineering designs for the construction of such a plant. On the basis of such designs, analysts predict that a plant equipped with CCS technology will cost more to build and to operate than will a conventionally equipped plant, for two main reasons:

- The equipment a CCS plant requires to capture and compress CO_2 is large, complex, and expensive; and
- Capturing and compressing CO_2 consumes a substantial fraction of the plant's total output. Consequently, to produce the same amount of electricity for customers, a plant with CCS capabilities has to be bigger than a plant without them.

According to CBO's analysis, average capital costs for a CCS-equipped plant would be 76 percent higher than those for a conventional plant: $3,070 per kilowatt of capacity compared with $1,740 per kilowatt. A CCSequipped plant would also be more expensive to operate than a non-CCS-equipped plant would be because it would have to burn more fuel during the process of capturing and compressing the CO_2. Those higher capital and operating costs would in turn make the electricity generated by newly constructed CCS-equipped plants more expensive than that generated by conventional coal-fired plants.

CBO analyzed five engineering studies of the estimated costs for building and operating a new coal-fired power plant that would include technology for capturing, transporting, and storing carbon dioxide.[12] (The appendix discusses those studies more fully. Calculating the estimated costs for the construction of new plants, rather than the costs for retrofitting existing ones, is more useful for comparing the various studies' findings.) The studies provided cost estimates for two types of generating facilities whose costs would be fairly representative of the industry's experience: subcritical pulverized coal plants (which employ the most common coal-based generating technology in use in the United States) and supercritical pulverized coal plants (which use a newer technology that generates electricity more efficiently). CBO's calculations excluded gasification electricity-generating plants and oxy-fuel combustion plants because very few commercial power plants now use those technologies and estimates of construction costs for such plants will be less reliable than those for plants using postcombustion methods.[13]

Each of the five studies measured capital costs for building a CCS-equipped facility, expressed as the cost per kilowatt of generating capacity. Each one also calculated the average cost of producing electricity at such a plant over its lifetime, including the cost of transporting and storing the captured CO_2, expressed in dollars per megawatt-hour.[14] CBO converted all costs to 2010 dollars and adjusted several of the assumptions underlying the studies to provide a common basis for comparing their findings. Those adjustments involved the cost of a coal-fired plant's financing, the cost of coal, and the operating capacity of a plant, among others (see the appendix for details).

On average, the difference in the cost of electricity produced with and without CCS technology, as reported in the studies that CBO reviewed, was estimated to be $45 per megawatt-hour—that is, $104 with CCS versus $59 without it. That difference, like the cost premium for constructing a CCS-equipped plant versus a conventionally equipped one, is 76 percent.

Reducing Cost Differentials

Experience in applying new technologies to coal-fired power plants indicates that the cost of generating electricity with such a technology may decline over time. For example, one recent study examined reductions in the costs of seven technologies relevant to power plants equipped with machinery for capturing carbon dioxide. As the technologies became more widely applied and equipment manufacturers and construction companies became more familiar with them, costs fell—a pattern of cost reduction called learning-by-doing. The study used the historical average reduction in costs for each component technology, weighted by the technology's share of a plant's total costs, to estimate likely future declines in costs for carbon capture technology. That analysis indicated that if 100 gigawatts of new CCS capacity was installed—either in the United States or abroad—the cost of producing electricity in a CCS-equipped coal-fired plant would drop by between 10 percent and 18 percent.[15]

The potential for gradual reductions in CCS-related costs is borne out by experience with other types of emission-control technologies in the electricity generation industry, but the process of reducing costs can be slow and sometimes requires additional funding for R&D activities or a substantial period spent using the new approach. Starting in the 1970s, as required by the Clean Air Amendments of 1970 and increasingly stringent regulations that were subsequently adopted, utilities reduced sulfur dioxide emissions from their smokestacks.

The 1990 amendments to the Clean Air Act added requirements for reducing emissions of nitrogen oxide. Yet despite power plants' long history of using technologies to capture sulfur dioxide emissions as well as funding from DOE for research, it took years to produce a technology that was efficient enough to meet the laws' more stringent requirements. As a result, costs dropped slowly. After 20 years of investment in such facilities, however, capital costs for the equipment had fallen by half, and the removal of sulfur dioxide from plants' emissions had become markedly more efficient.[16] To reduce costs by that much, the industry had put in place sulfur dioxide emission controls on power plants worldwide that together produced almost 200 GW of electricity.

Projections of the cost of CCS technology rest on engineering estimates and the learning-curve models, but two major sources of uncertainty characterize the findings from such analyses. First, the costs for building a CCSequipped plant could be substantially greater than the initial estimates that the engineering studies present. The builders of a precombustion CCS plant in Mississippi, for example, recently announced that they expected an increase of $366 million in the project's previously estimated cost of $2.4 billion. Second, costs do not always decline as smoothly as learning curves suggest.

Table 3. CBO's Illustrative Calculations of the Estimated Reduction in the Cost of Electricity from CCS-Equipped Plants

(2010 dollars per megawatt-hour)

	Levelized Cost of Electricity[a]
	Costs When the First CCS Plant Goes into Operation
Initial CCS Plant Coal-Fired Plant Without CCS	104
Coal-Fired Plant Without CCS	59
CCS Cost Differential (Percent)	76
	Costs After Investment in 210 Gigawatts of CCS Capacity Worldwide
CCS Plant After 210 Gigawatts of Worldwide Investment	74
Coal-Fired Plant Without CCS[b]	55
CCS Cost Differential (Percent)	35
Memorandum: Cost Reduction for CCS Plant per 100 Gigawatts of New Investment (Percent)	14

Source: Congressional Budget Office.

Notes: In this analysis, a power plant's electricity-generating capacity is measured in kilowatts; the electrical power generated by that capacity is measured in megawatt-hours. During a full hour of a plant's operation, 1 kilowatt of capacity produces 1 kilowatt-hour of electricity; 1,000 kilowatt-hours equals 1 megawatt-hour.

CCS = carbon capture and storage (the set of processes and technologies that separate carbon dioxide, or CO_2, from other gases generated when a fossil fuel is burned; compress the CO_2 into a fluid; and then transport it to an underground location for storage).

[a] The levelized cost of electricity is the average cost over a plant's lifetime for producing a megawatt-hour of electricity, taking into account such factors as the cost of building the plant, debt service, the return on equity investment, tax rates, fuel costs, operating expenses, and the plant's electricity-generating capacity.

[b] The cost of producing electricity at conventional coal-fired plants would also be expected to decline as more experience is gained in building such plants. However, the rate of that decline would probably be much slower than the decline in costs for CCS-equipped plants.

In many cases, the first version of a new technology proves inadequate and requires redesigning, leading to costs for subsequent plants that are higher than those for the initial plant—although eventually, those higher costs decline as experience is gained. That pattern has characterized the introduction and adoption of some earlier technologies used to reduce emissions from coal-fired power plants; indeed, one study of electricity prices after the introduction of new emission-related equipment reported that initial costs increased in the majority of cases before such costs began to decline.[17]

How Much New Capacity Using CCS Technology Must Be Built to Reduce the Cost of Generating Electricity in a CCS-Equipped Plant?

According to CBO's estimates, roughly 210 GW of generation capacity in the form of new CCS-equipped coal-fired power plants would be required to achieve DOE's cost reduction goal for the technology (see Table 3). (That capacity is equal to about two-thirds of the total capacity—317 GW—of coal-fired generating plants operating in the United States.)[18] CBO based its calculations on the initial difference in costs that it derived from its review of engineering design studies for new CCS-equipped plants and its analysis of learning-curve studies of earlier emission-control technologies. Specifically, CBO used the following assumptions in its calculations:

- The average cost of electricity from a new coal-fired power plant without CCS capabilities is $59 per megawatt-hour, and the average cost of electricity megawatt-hour, or 76 percent more (see the appendix for details).
- The cost of building and operating a CCS-equipped plant will fall by 14 percent for every 100 GW of additional capacity that is built. That assumption is the midpoint of the range of learning curves discussed earlier (10 percent to 18 percent).
- The cost of producing electricity in a new conventional coal-fired power plant will fall by 6 percent during the time required to add 210 GW of CCSequipped capacity. According to EIA's forecasts, coal-fired generation capacity worldwide will increase by 66 percent between 2007 and 2035, an expansion that could provide opportunities for improvements in constructing and operating conventional as well as CCS-equipped plants.[19] If the development of CCS proceeds slowly, the reductions in the cost of the competing conventional capacity could be greater than CBO has assumed.

Because the amount of new CCS-equipped capacity needed to achieve DOE's cost reduction goal would vary under assumptions different from those above, CBO performed several analyses to explore those effects. Specific estimates of the amount of necessary additional capacity varied widely, but they were all substantial. For example, if the initial cost of a CCS-equipped plant turned out to be a third higher than the average of the estimates on which CBO based its analysis—so that such a plant cost twice as much as a new conventional coal-fired plant— 310 GW of additional capacity (rather than 210 GW) would have to be built to meet DOE's goal. By contrast, if DOE's R&D program proved fruitful and reduced the cost of building a CCS-equipped plant to only 66 percent more than the cost of building a new conventional plant, then 170 GW of additional capacity would be sufficient to meet the goal. Overall, those estimates ranged from an amount at the low end roughly equal to half the United States' current coal-fired generating capacity to an amount at the high end roughly equal to all of the country's capacity.

Changing the pace of learning would also affect CBO's illustrative calculations. Those calculations used the midpoint (14 percent) of the learning-curve range of reductions in costs per 100 GW of new capacity (10 percent to 18 percent). If, instead, learning occurred more slowly and the cost of producing electricity at CCSequipped plants fell by 10 percent for each 100 GW of new capacity, it would take 330 GW of additional capacity to reduce costs to the requisite level. Alternatively, if learning occurred more quickly and electricity production costs declined by 18 percent per 100 GW of new capacity, then 155 GW would be needed.

The cost of generating electricity at plants equipped with CCS technology would also be expected to decline over time if existing plants were retrofitted to capture CO_2 emissions. That is, the learning necessary to reduce costs would be similar regardless of whether the technology was installed in a new facility or an existing one. However, the cost of using CCS would generally be less at a new plant because the equipment and processes could be standardized, whereas at an existing plant, the technology would have to be adapted to the facilities that were already in place, which is usually a more expensive approach. Nevertheless, the higher costs of the CCS equipment at existing facilities might be more than offset by avoiding the array of construction costs that a new facility would entail. The decision about retrofitting existing plants rather than building new ones is complex and would be affected by many factors other than the cost of the CCS equipment.

How Much Construction of New Electricity-Generating Capacity Is Projected to Occur in the Near Future?

Given the economics of producing electricity with CCS technology, there seems to be little likelihood of substantial investment in that technology in the near future, either in the United States or elsewhere, particularly if no laws or agreements are in place to limit CO_2 emissions.

Current projections indicate that the United States is unlikely to need an additional 210 GW of coal-fired generating capacity in the near future. (One reason is that in recent years, natural gas has accounted for a greater share of electricity generation than it did in the past; see Box 1.) EIA has projected that the average annual growth in capacity in the entire electric power sector will be 0.4 percent per year through 2035, for a total addition of 187 GW of new capacity.[20] So even if all new generation capacity built by electric power companies in the United States through that year was devoted to CCS, it would probably be insufficient to meet DOE's cost reduction targets.

But currently, little incentive exists for investments in CCS capacity. That might change if the United States implemented policies to reduce CO_2 emissions; then, the domestic stock of electricity generation plants would turn over more rapidly, and the potential for new investment in CCS capacity in the United States—both in new plants and in existing plants that might be retrofitted— would be strengthened. If DOE met its goal of reducing the cost premium for CCS-generated electricity to no more than 35 percent while capturing 90 percent of the associated CO_2 emissions, utilities and others that might build CCS-equipped plants would still operate at a cost disadvantage of $19 per megawatt-hour ($74 versus $55; see Table 3 on page 9). However, if policymakers also imposed a price (say, as a tax or as part of a cap-and-trade system) of $20 per metric ton of CO_2 emissions, then, in CBO's estimation, that cost disadvantage would be eliminated.[21]

Investors and consumers that faced a price on CO_2 emissions could choose from among several approaches for reducing them: for example, coal-fired generation with CCS, conservation, or the use of other sources of energy, such as natural gas (with or without CCS), nuclear power, wind, biomass, or other renewables. Overall, investors would probably choose a mix of all of those types of investments, which would be determined in large part by the relative prices of the different options for reducing emissions. In one study that examined how the electricity industry could reduce its CO_2 emissions if policymakers imposed a price on them, the role played by CCS in the future varied widely—ranging from accounting for roughly half of all electricity generation to of 187 GW of new capacity.[20] So even if all new generation capacity built by electric power companies in the United States through that year was devoted to CCS, it would probably be insufficient to meet DOE's cost reduction targets.

But currently, little incentive exists for investments in CCS capacity. That might change if the United States implemented policies to reduce CO_2 emissions; then, the domestic stock of electricity generation plants would turn over more rapidly, and the potential for new investment in CCS capacity in the United States—both in new plants and in existing plants that might be retrofitted— would be strengthened. If DOE met its goal of reducing the cost premium for CCS-generated electricity to no more than 35 percent while capturing 90 percent of the associated CO_2 emissions, utilities and others that might build CCS-equipped plants would still operate at a cost disadvantage of $19 per megawatt-hour ($74 versus $55; see Table 3 on page 9). However, if policymakers also imposed a price (say, as a tax or as part of a cap-and-trade system) of $20 per metric ton of CO_2 emissions, then, in CBO's estimation, that cost disadvantage would be eliminated.[21]

Investors and consumers that faced a price on CO_2 emissions could choose from among several approaches for reducing them: for example, coal-fired generation with CCS, conservation, or the use of other sources of energy, such as natural gas (with or without CCS), nuclear power, wind, biomass, or other renewables. Overall, investors would probably choose a mix of all of those types of investments, which would be determined in large part by the relative prices of the different options for reducing emissions. In one study that examined how the electricity industry could reduce its CO_2 emissions if policymakers imposed a price on them, the role played by CCS in the future varied widely—ranging from accounting for roughly half of all electricity generation to of 187 GW of new capacity.[20] So even if all new generation capacity built by electric power companies in the United States through that year was devoted to CCS, it would probably be insufficient to meet DOE's cost reduction targets.

But currently, little incentive exists for investments in CCS capacity. That might change if the United States implemented policies to reduce CO_2 emissions; then, the domestic stock of electricity generation plants would turn over more rapidly, and the potential for new investment in CCS capacity in the United States—both in new plants and in existing plants that might be retrofitted— would be strengthened. If DOE met its goal of reducing the cost premium for CCS-generated electricity to no more than 35 percent while capturing 90 percent of the associated CO_2 emissions, utilities and others that might build CCS-equipped plants would still operate at a cost disadvantage of $19 per megawatt-hour ($74 versus $55; see Table 3 on page 9). However, if policymakers also imposed a price (say, as a tax or as part of a cap-and-trade system) of $20 per metric ton of CO_2 emissions, then, in CBO's estimation, that cost disadvantage would be eliminated.[21]

Investors and consumers that faced a price on CO_2 emissions could choose from among several approaches for reducing them: for example, coal-fired generation with CCS, conservation, or the use of other sources of energy, such as natural gas (with or without CCS), nuclear power, wind, biomass, or other renewables. Overall, investors would probably choose a mix of all of those types of investments, which would be determined in large part by the relative prices of the different options for reducing emissions. In one study that examined how the electricity industry could reduce its CO_2 emissions if policymakers imposed a price on them, the role played by CCS in the future varied widely—ranging from accounting for roughly half of all electricity generation to accounting for none—depending on the relative size of the cost premiums for CCS- and nuclear-generated power compared with power generated in conventional coal-fired plants.[22]

Another possibility is that the United States could benefit from knowledge about CCS gained through experience in other countries. In contrast to EIA's forecasts for the growth of capacity in the United States, its outlook for the rest of the world suggests that the needed investment of 210 GW in CCS-equipped capacity could be accommodated easily. By EIA's estimates, worldwide coal-fired generating capacity will grow by more than 940 GW during the 2007–2035 period, rising from 1,425 GW to 2,366 GW.[23] About 80 percent of that increase would occur in other countries, where, unlike projected demand in the United States, the demand for electricity is growing rapidly; China and India in particular are poised to experience increases. Thus, investments in CCS-equipped generation plants abroad could provide enough industrywide learning to reduce costs for similar plants built in the United States. (In the same way, investments in plants in the United States would probably provide some experiential benefits to foreign generators.)

The potential for reciprocal technology transfer between the United States and other countries is limited in the short term, however, because of differences in the stages of various countries' development of the technology. China, for example, is currently focused on research and development, together with small-scale demonstration projects; funding for large-scale demonstration projects is limited.[24] Large-scale CCS projects are being built mainly in North America and Europe, where capacity is growing much more slowly.

Over the longer term, shifting growth in the world's coal-based electricity-generating capacity toward investment in CCS would require international agreements on substantial reductions in CO_2 emissions.

Box 1. Natural Gas-Fired Electricity Generation and the Development of Carbon Capture and Storage Technologies

Most public and private efforts to develop technologies for capturing, compressing, transporting, and storing carbon dioxide (CO_2) emitted during electricity generation have focused on coal-fired plants rather than on natural gas-fired plants. Until recently, natural gas has been more expensive than coal as a power source and thus has been used less; in addition, coal-fired power plants produce more electricity than their natural gas-fired counterparts, and coal combustion produces far more emissions of CO_2 per unit of electricity than does the burning of natural gas. But with lower prices for natural gas and its increasing use for electricity generation, carbon capture and storage (CCS) technology could also be developed for use in natural gas-fired power plants. Nevertheless, because demonstration and pilot projects to construct facilities equipped with CCS technology take years to plan and fund, and because the potential benefits of CCS are greater for coal-fired plants, the use of CCS at coal-fired facilities will probably remain at the forefront of the technology's development for at least the next few years.

Increased Use of Natural Gas in Electricity Generation

The role that natural gas plays in the production of electricity in the United States has been shaped by several factors. Capital costs for electricity-producing turbines powered by natural gas are lower than those for coal-fired or nuclear-powered generators; thus, by using natural gas, utilities can increase their capacity to generate electricity while incurring lower capital costs than would be necessary if those other power sources were used. However, because natural gas generators have, until the past few years, been more expensive to operate than coal-fired or nuclear-powered facilities, they have mainly been used at times of peak demand.

Now, for several reasons, many analysts expect utilities to use natural gas more regularly to satisfy normal rather than just peak demand; those reasons include the fuel's relatively low price; the cost of addressing environmental effects other than the buildup of greenhouse gases, such as emissions of mercury; and uncertainty about the cost of complying with possible future regulation of greenhouse gas emissions.[1] A striking indication of that trend is that since 2000, the amount of electricity generated in coal-fired power plants has fallen by 7 percent and the amount of electricity generated in natural gas-fired plants has grown by 64 percent.[2]

That shift toward using natural gas in nonpeak times will undercut the possible need for coal-fired CCS plants to meet environmental goals. Natural gas facilities emit roughly half the CO_2 that a similarly sized coal-fired plant emits.[3] If natural gas prices remained low or the regulation of greenhouse gases became more stringent, natural gas-fired plants could prove an increasingly viable alternative to coal-fired plants. In fact, the Environmental Protection Agency (EPA) is considering regulations to limit emissions of CO_2 to a level below those produced by a typical coal-fired plant that is not equipped with CCS technology.

Carbon Capture and Storage Technology for Natural Gas-Fired Power Plants

CCS technology could be adopted at plants that use natural gas rather than coal for electricity generation. In fact, the cost of using CCS at natural gas-fired plants would probably be less than the cost of using it at coal-fired plants. In particular, although much of the capture and compression equipment integral to the CCS approach is the same for both types of plants (and the transportation and storage facilities are identical), natural gas-fired plants would require less equipment because they produce fewer CO_2 emissions.

Similarly, developers of CCS-capable coal-fired facilities face additional challenges to ensure that the new emission-reduction technology is compatible with the plant's coal-handling and exhaust equipment, some of which limits the emissions of other pollutants.

However, firms today have no incentive to install CCS technology in natural gas-fired plants. It is not economically viable now, and no policies are in place to encourage utilities to purchase the additional equipment and incur the higher costs for producing electricity that CCS technology currently entails.

If EPA adopted regulations for reducing greenhouse gases or lawmakers enacted policies to restrict emissions from power plants (such as imposing a price on CO_2 emissions or subsidizing the production of electricity at CCS-equipped electricity generation plants), then natural gas-fired facilities might have an incentive to use CCS to meet such requirements. At present, though, regulatory action to curb CO_2 emissions is more likely to shift electricity production from coal to natural gas (without CCS) and other low-emission fuels, such as biomass, rather than to CCS-capable plants.[4]

[1] Susan Tierney, Why Coal Plants Retire: Power Market Fundamentals as of 2012 (Analysis Group, February 16, 2012), pp. 3–4.

[2] Department of Energy, Energy Information Administration, Electric Power Annual 2010 (November 2011), Table ES1, www.eia.gov/electricity

[3]. Utilities that generate electricity by burning natural gas also use that process as an inexpensive way to reduce emissions of mercury and other nongreenhouse gases.

[4]. Nathan Richardson, Art Fraas, and Dallas Burtraw, Greenhouse Gas Regulation Under the Clean Air Act: Structure, Effects, and Implications of a Knowable Pathway, RFF DP 10-23 (Resources for the Future, April 2010), pp. 43–45, www.rff.org/rff/Documents/RFF-DP-10-23.pdf.

Policy Options

In CBO's view, current policies are unlikely to achieve the goal of reducing the additional costs for producing electricity with CCS technology to 35 percent more than the cost of producing electricity without CCS. DOE's present funding for CCS would allow the United States to build only a small number of demonstration plants, which are unlikely to be sufficient to reduce costs through the learning process described earlier. If DOE adhered to its current plan, it would continue to support the R&D and demonstration programs for which the American Recovery and Reinvestment Act provided funding of $3.4 billion, and it would continue to seek annual appropriations of $300 million to $400 million for related efforts.

However, unless lawmakers substantially increased support for CCS, probably well beyond even those amounts, federal funding would be likely to contribute only a little to reducing the costs of CCS-equipped coal plants after the initial demonstration projects for the technology had ended. Most investment in electric utilities comes from the private sector. As CBO's illustrative calculations suggest, the amount of current federal spending is small relative to the magnitude of the investment necessary to make CCS-equipped plants economically competitive, and DOE's current activities are unlikely to provide the amount of learning that would drive down the technology's costs. Rather, reductions would have to be spurred by the activities of investors and the efforts of utilities and their customers.

To encourage investment and deployment, additional incentives would be needed, many analysts say; without them, the returns on investment in CCS plants beyond the demonstration stage would be too small to attract investors. Yet the history of other emission-control technologies suggests that the ability of widespread deployment alone to reduce costs is limited. Even with broad deployment, such earlier technologies as the removal of sulfur dioxide from utilities' emissions have required decades of experience and extensive research and development—as well as a substantial amount of investment—before costs were reduced.[25]

The decisions facing policymakers with regard to support for CCS center on whether current federal technology programs, which are mainly devoted to reducing the costs of the capture portion of the process, should continue as they are currently structured. Some alternatives include shifting the focus from demonstration projects to research and development, adopting policies that encourage private investment in CCS, or reducing or eliminating support for CCS.

Shift DOE's Focus from Demonstration Projects to Research and Development

One option for increasing the effectiveness of federal spending on CCS technology would be to limit that support to research and development and withdraw it from more-costly demonstration projects. Concentrating federal resources on R&D would focus DOE's efforts on activities for which the rationale for spending by the federal government is the strongest—that is, in bringing scientists and engineers together to perform research that is removed from specific commercial applications. R&D is also an area in which the federal record of success is long, compared with many failed federal attempts to commercialize earlier fossil energy technologies.[26]

One variation of that option would be for the federal government to collaborate with governments of other countries in developing CCS technology. DOE, sometimes in cooperation with the U.S. Agency for International Development, supports many activities through the United Nations, the International Energy Agency, and other multilateral groups to promote the transfer of CCS technology internationally. DOE is also promoting adoption of U.S.-developed environmental technology in China as well as the use of U.S.-developed clean coal technologies in other countries. In other areas—for example, research in particle physics and fusion energy— DOE collaborates with foreign partners to pay for large projects, which in many cases are not built on U.S. soil. Such collaboration might be a model that could be applied to CCS-equipped power plants. However, without international agreements on reducing CO_2 emissions, such new projects are likely to be developed only slowly, if at all.

Adopt Policies That Encourage Private Investment in CCS-Equipped Plants

The private sector would have a greater incentive to invest in CCS technology if the federal government adopted policies that in some way offset the higher cost of generating electricity in coal-fired plants equipped with the technology. Currently, the price of electricity reflects the cost of producing it but not the cost of the damage that could be expected from climate change caused by emissions of greenhouse gases. The United States could adopt other policies that would incorporate the cost of that damage in the price of electricity. For example, imposing a tax on CO_2 emissions or adopting a cap-andtrade program that would limit emissions would cause the price of electricity from conventional coal-fired plants to increase. If it did, that rise would encourage investment in technologies like CCS that reduced emissions. However, even with such a policy in place, investors might choose other approaches for reducing CO_2 emissions because of the high cost of CCS.

Similarly, some analysts have suggested that instead of subsidizing the development and construction of CCSequipped plants, the federal government could subsidize the electricity that the plants produce—for example, by offering to pay any utility using CCS technology a fixed amount for each ton of CO_2 that it captured and stored. Such a policy would avoid having DOE choose individual projects to support and could focus federal efforts on, for example, determining whether such plants were indeed producing electricity in a process that reduced emissions of CO_2. Currently, electric utilities can receive a tax credit for using CCS technology to generate electricity. However, the credit does not seem to have encouraged private investors to equip plants with CCS— either because the credit has been too small or because the credit does not provide an incentive for firms that do not pay taxes.

But subsidizing the electricity that CCS-equipped plants produce would be another form of federal investment in a technology that might never prove to be cost-effective. Moreover, lawmakers have already committed large amounts of money and many years to the construction of the next generation of CCS-equipped plants, and DOE has signed agreements to help fund the construction of several of them.

Reduce or Eliminate DOE's Support for CCS

Given the limits on DOE's ability to lower the costs of CCS through its currently planned activities, lawmakers could substantially reduce or discontinue funding for both developing and demonstrating the technology. If little coal-fired generation capacity was being built in the United States, lawmakers might decide that the development of technologies such as CCS would have little effect on either reducing CO_2 emissions or preserving the nation's ability to use coal-fired power plants in the future. Moreover, even if DOE's cost reduction target was attained, coal-fired plants with CCS would not be competitive with plants that lacked the technology unless policies were adopted that imposed costs on carbon emissions.

Scaling back or eliminating the CCS programs would reduce the need for future annual appropriations for those activities. Moreover, eliminating larger-scale technology demonstration projects would reduce DOE's involvement in fields in which the agency has a mixed track record and in which U.S. industry is generally not poised to follow up with subsequent investment.

An option that would reduce or discontinue support for CCS would not necessarily apply to the funding already provided for demonstration projects, however. Much of that money has been obligated (that is, legally committed for some purpose that will result in outlays) but not yet spent, and because of the CCS-equipped demonstration plants that have been canceled or put on hold, a great deal of it may never be spent. The eventual disposition of those obligated but unspent funds is currently unknown. Because DOE has signed agreements with several private investors to help pay for the five large-scale demonstration plants that are still being built or that are planned to be built, spending for CCS could not be eliminated immediately. In addition, because of existing agreements, DOE might bear some shutdown costs if its support of those plants was terminated or reduced.

APPENDIX:
DEVELOPING A COMMON BASIS FOR COMPARING ENGINEERING COST ESTIMATES

The set of facilities and processes known as carbon capture and storage (CCS) technology would allow electric utilities that install the technology to capture emissions of carbon dioxide and store them as a liquid underground. The Congressional Budget Office (CBO) compared the costs for building and operating new coal-fired power plants with and without CCS technology. That comparison is based on estimates from five engineering studies, published from 2005 to 2010, of the cost of equipping new electricity-generating plants with CCS technology. Those studies consist of engineering designs and their associated costs for what are considered reasonable plans for constructing and operating such a plant. (No full-scale CCS-capable facility has as yet been built.) CBO used two principal measures of costs in its calculations:

- Total plant construction costs—the costs for building a new power plant per kilowatt of net electrical output that the plant can produce;[1] and
- The levelized cost of electricity—the average cost of producing a megawatt-hour of power over the lifetime of a plant, taking into account such factors as the costs of building, financing, and operating the facility.

On the basis of the studies' findings, CBO concluded that the added cost of constructing and operating a CCS-equipped coal-fired plant rather than a conventional coal-fired power plant would average about 76 percent for both total plant costs and for the levelized cost of electricity (see Table A-1).

For its analysis, CBO converted all costs to 2010 dollars and adjusted several of the assumptions underlying the studies to provide a common basis for comparing their findings. CBO's adjustments included the following:

- The engineering studies used a nominal *fixed-charge factor* to take into account the cost of financing for the plant's construction, calculated on an annual basis. The nominal fixed-charge factor includes the cost of funds used for construction plus payments toward the principal of any loans, calculated over the lifetime of the plant (which is assumed to be 30 years). The cost of funds represents a weighted cost of the debt and equity financing used in building the plant. (The cost of debt is the interest paid on any bonds that are issued, and the cost of the equity financing is the return to investors in a power plant; both costs are adjusted to take taxes into account.) The fixed-charge factors that the studies used ranged from 10.5 percent to 15.1 percent, a variation that reflects the general rise in interest rates leading up to 2008 and the subsequent drop in rates after 2009. To compare the studies on an equal basis, CBO used a nominal fixed-charge factor of 10.5 percent to reflect the lower interest rates in recent years.

- Assumptions in the engineering studies about the price of coal over the 30-year lifetime of a plant ranged from $1.50 to $2.47 per million British thermal units (Btu) of heat produced. The price of coal assumed in CBO's analysis was $2.23 per million Btu. CBO derived that estimate by averaging forecasts for 2010 to 2035 of the price of coal used for electricity generation (developed by the Department of Energy's Energy Information Administration) and then converting that number to 2010 dollars.

- Operating capacity factors in the engineering studies ranged from 80 percent to 85 percent. (The capacity factor measures how much of a plant's total capacity for generating electricity is used over the course of a year.) A larger capacity factor will reduce the levelized cost of electricity because fixed costs can be spread over the production of more energy. Because four of the five studies used 85 percent as the capacity factor, CBO adjusted the capacity factor of the other study to equal 85 percent.

- To account for the *increase in power plant construction costs* from 2005 to 2010, CBO adjusted the studies' results to reflect costs in 2010. From 2005 to 2008, the nominal costs for building a new power plant rose by about 40 percent; between 2008 and 2010, however, costs fell by about 7 percent, making the estimates of costs for the beginning of the period quite different from those for the end.

- Similarly, to account for the increase in nominal *operating expenses*—that is, those not related to CCS operations—in the various years of the studies, CBO adjusted the studies' estimates of nonfuel operating expenses to reflect 2010 costs. The operating costs of a power plant increased by 20 percent from 2005 to 2008 before declining by about 9 percent in 2009.

- All of the studies except the one by the Massachusetts Institute of Technology (MIT) included a *cost for storing and transporting the captured carbon dioxide*. To account for such a cost and thus compare the MIT study with the others on an equal-cost basis, CBO added $10 for each short ton (2,000 pounds) of captured carbon dioxide to the MIT study's estimate of the levelized cost of electricity.[2]

The adjusted estimates of total plant costs for non-CCSequipped plants ranged from about $1,600 to $1,900 per kilowatt of net electrical output (see Table A-2); the adjusted estimates for CCS-equipped plants ranged from about $2,800 to $3,500 per kilowatt. After the adjustments, the estimated cost premium for building a CCS plant ranged from 61 percent to 89 percent in the studies. The adjusted estimates of the levelized cost of electricity for non-CCS plants ranged from $53 to $66 per megawatt-hour, and the adjusted estimates for CCS-equipped plants ranged from $95 to $112 per megawatt-hour. The latter figures represent a premium of 70 percent to 89 percent in the levelized cost of electricity from a CCS plant.

Table A-1. Estimates from Engineering Studies of Total Plant and Levelized Electricity Costs for New Coal-Fired Power Plants With and Without CCS Technology

	Total Plant Construction Costs[a] (Dollars)			Levelized Cost of Electricity[b] (Dollars)	
	Supercritical Pulverized Coal Plants[c]	Subcritical Pulverized Coal Plants[c]	Fixed-Charge Factor[d] (Percent)	Supercritical Cost of Coal Pulverized (Dollars per MMBtu)Coal Plants[c]	Subcritical Pulverized Coal Plants[c]
	Carnegie Mellon University (2009 dollars)				
Without CCS	1,770	1,693	10.5	2.3	55.4	55.6
With CCS	3,205	3,202	10.5	2.3	96.2	99.9
Premium for CCS (Percent)	81	89	0	n.a.	74	80
	Electric Power Research Institute (2006 dollars)				
Without CCS	1,763	e	11.7	1.5	53.2	e
With CCS	2,930	e	11.7	1.5	92.8	e
Premium for CCS (Percent)	66	e	0	n.a.	74	e
	Global Carbon Capture and Storage Institute (2010 dollars)				

Table A-1. (Continued)

	Total Plant Construction Costs[a] (Dollars)			Levelized Cost of Electricity[b] (Dollars)		
Without CCS	1,919	[e]	12.1	2.5	76.0	[e]
With CCS	3,464	[e]	12.1	2.5	131.0	[e]
Premium for CCS (Percent)	81	[e]	0	n.a.	72	[e]
	Massachusetts Institute of Technology (2005 dollars)					
Without CCS	1,330	1,280	15.1	1.5	47.8	48.4
With CCS	2,140	2,230	15.1	1.5	76.9[f]	81.6[f]
Premium for CCS (Percent)	61	74	0	n.a.	61	69
	National Energy Technology Laboratory (2007 dollars)					
Without CCS	1,647	1,622	13.4	1.6	74.7	75.3
With CCS	2,913	2,942	14.0	1.6	135.2	139.0
Premium for CCS (Percent)	77	81	4	n.a.	81	85

Source: Congressional Budget Office based on Massachusetts Institute of Technology, The Future of Coal (2007), web The_Future_of_Coal.pdf; Electric Power Research Institute, Updated Cost and Performance Estimates for Clean Coal Technologies Including CO2 Capture—2006, EPRI Report 1013355 (March 2007), http://mydocs.epri.com/docs/public/ 000000000001013355.pdf; National Energy Technology Laboratory, Cost and Performance Baseline for Fossil Energy Plants, vol. 1, Bituminous Coal and Natural Gas to Electricity, DOE/NETL-2010/1397 (November 2010), www.netl.doe.gov/technologies/ coalpower/ewr/ co2/SystemsAnalysis.html; Carnegie Melon University, Integrated Environmental Control Model, version 6.24 (May 2010), www.cmu.edu/epp/ iecm/iecm_dl.html; and Global CCS Institute, Economic Assessment of Carbon Capture and Storage Technology: 2011 Update (2011), www.globalccsinstitute. com/publications/ economic-assessment-carbon-capture-andstorage-technologies-2011-update.

Notes: In this analysis, a power plant's electricity-generating capacity is measured in kilowatts; the electrical power generated by that capacity is measured in megawatt-hours. During a full hour of a plant's operation, 1 kilowatt of capacity produces 1 kilowatt-hour of electricity; 1,000 kilowatt-hours equals 1 megawatt-hour.

All of the studies except the one conducted by the Electric Power Research Institute (EPRI) incorporated the assumption that over a year, a plant would operate 85 percent of the time. In contrast, the EPRI study incorporated an assumed annual rate of operation of 80 percent.

MMBtu = 1 million British thermal units (1 BBtu is the amount of energy required to raise the temperature of 1 pound of water by 1 degree Fahrenheit under certain controlled conditions); CCS = carbon capture and storage (the set of processes and technologies that separate carbon dioxide, or CO2, from other gases generated when a fossil fuel is burned; compress the CO2 into a liquid; and then transport it to an underground location for storage); n.a. = not applicable.

[a] The costs for building a new power plant for each kilowatt of net electrical output that the plant can produce.

[b] The levelized cost of electricity is the average cost over a power plant's lifetime for producing 1 megawatt-hour of electricity, taking into account such factors as the cost of building, financing, and operating the plant.

[c] "Subcritical" and "supercritical" refer to the efficiency of electricity production from pulverized coal—that is, the percentage of the fuel's potential energy that is actually converted to electricity. The efficiency of electricity production using subcritical pulverized coal is about 37 percent; the efficiency of electricity production using supercritical pulverized coal, which uses higher steam temperatures and pressures during generation, is about 40 percent.

[d] Broadly speaking, the cost of the funds required to build the plant, including the rate of interest on any debt and the rate of return on equity investment, adjusted to take federal, state, and property taxes into account.

[e] EPRI and the Global Carbon Capture and Storage Institute did not estimate costs for a subcritical pulverized coal plant.

[f] Does not include the cost of storing and transporting the captured CO_2.

Table A-2.

CBO's Adjusted Estimates of Total Plant and Levelized Electricity Costs for New Coal-Fired Power Plants With and Without CCS Technology (2010 dollars)

	Total Plant Construction Costs[a]		Levelized Cost of Electricity[b]	
	Supercritical Pulverized Coal Plants[c]	Subcritical Pulverized Coal Plants[c]	Supercritical Pulverized Coal Plants[c]	Subcritical Pulverized Coal Plants[c]
	Carnegie Mellon University			
Without CCS	1,788	1,710	55.9	56.0
With CCS	3,237	3,234	97.3	100.8
Premium for CCS (Percent)	81	89	74	80
	Electric Power Research Institute			
Without CCS	1,888	[d]	65.5	[d]

Table A-2. (Continued)

	Total Plant Construction Costs[a]		Levelized Cost of Electricity[b]	
	Supercritical Pulverized Coal Plants[c]	Subcritical Pulverized Coal Plants[c]	Supercritical Pulverized Coal Plants[c]	Subcritical Pulverized Coal Plants[c]
With CCS	3,138	D	111.5	d
Premium for CCS (Percent)	66	d	70	d
		Global Carbon Capture and Storage Institute		
Without CCS	1,919	d	57.4	d
With CCS	3,464	d	101.8	d
Premium for CCS (Percent)	81	d	77	d
		Massachusetts Institute of Technology		
Without CCS	1,734	1,669	53.1	54.6
With CCS	2,790	2,907	95.4	103.3
Premium for CCS (Percent)	61	74	79	89
		National Energy Technology Laboratory		
Without CCS	1,637	1,612	63.2	64.0
With CCS	2,895	2,924	107.7	111.3
Premium for CCS (Percent)	77	81	71	74

Source: Congressional Budget Office based on Massachusetts Institute of Technology, The Future of Coal (2007), web The_Future_of_Coal.pdf; Electric Power Research Institute, Updated Cost and Performance Estimates for Clean Coal Technologies Including CO2 Capture—2006, EPRI Report 1013355 (March 2007), http://mydocs.epri.com/docs/public/ 000000000001013355.pdf; National Energy Technology Laboratory, Cost and Performance Baseline for Fossil Energy Plants, vol. 1, Bituminous Coal and Natural Gas to Electricity, DOF/NETL-2010/1397 (November 2010), www.netl.doe.gov/technologies/ coalpower/ ewr/co2/SystemsAnalysis.html; Carnegie Melon University, Integrated Environmental Control Model, version 6.24 (May 2010), www.cmu.edu/epp/ iecm/iecm_dl.html; and Global CCS Institute, Economic Assessment of Carbon Capture and Storage Technology: 2011 Update (2011), www.globalccsinstitute. com/publications/economic-assessment-carbon-capture-andstorage-technologies-2011-update.

Notes: CBO adjusted the results of the studies to provide a basis for comparing them, using common assumptions about the price of coal, a fixed-charge factor, and capacity utilization—that is, how much of the time during a year a plant would be operating. (Broadly speaking, the fixed-charged factor is the cost of the funds required to build the plant, including the rate of interest on any debt and the rate of return on equity investment, adjusted to take federal, state, and property taxes into account.) In addition, CBO converted all costs into 2010 dollars. In this analysis, a power plant's electricity-generating capacity is measured in kilowatts; the electrical power generated by that capacity is measured in megawatt-hours. During a full hour of a plant's operation, 1 kilowatt of capacity produces 1 kilowatt-hour of electricity; 1,000 kilowatt-hours equals 1 megawatt-hour.

CCS = carbon capture and storage (the set of processes and technologies that separate carbon dioxide, or CO_2, from other gases generated when a fossil fuel is burned; compress the CO_2 into a liquid; and then transport it to an underground location for storage).

[a] The costs for building a new power plant for each kilowatt of net electrical output that the plant can produce.

[b] The levelized cost of electricity is the average cost over a power plant's lifetime for producing 1 megawatt-hour of electricity, taking into account such factors as the cost of building, financing, and operating the plant.

[c] "Subcritical" and "supercritical" refer to the efficiency of electricity production from the pulverized coal—that is, the percentage of the fuel's potential energy that is actually converted to electricity. The efficiency of electricity production using subcritical pulverized coal is about 37 percent; the efficiency of supercritical pulverized coal, which uses higher steam temperatures and pressures during generation, is greater, at about 40 percent.

[d] The Electric Power Research Institute and the Global Carbon Capture and Storage Institute did not estimate costs for a subcritical pulverized coal plant.

End Notes

[1]. Electricity production and emission data are from, respectively, Department of Energy, Energy Information Administration, "Coal's Share of Total U.S. Electricity Generation Falls Below 40% in November and December," Today in Energy (March 9, m2012), www.eia.gov /todayinenergy/detail.cfm?id=5331&src=email, and Annual Energy Review 2010, DOE/EIA-0384(2010) (October 2011), Table 11.2, p. 317, www.eia.gov/ totalenergy/ data/annual/pdf/aer.pdf.

[2]. For more information, see Congressional Budget Office, The Economics of Climate Change: A Primer (April 2003), and Potential Impacts of Climate Change in the United States (May 2009).

[3]. For additional discussion, see Congressional Budget Office, How Policies to Reduce Greenhouse Gas Emissions Could Affect Employment (May 2010).

[4]. Department of Energy, Energy Information Administration, Annual Energy Outlook 2012 (June 2012), p. 95, www.eia.gov/forecasts/aeo/.

[5]. For estimates of the quantities of CO2 used for such efforts, see Department of Energy, Office of Fossil Energy, Report of the Interagency Task Force on Carbon Capture and Storage (August 2010), p. D-1, www.fe.doe.gov/programs/sequestration/ccs_task_force.html.

[6]. Ibid., p. 39. For more information, see Congressional Budget Office, The Potential for Carbon Sequestration in the United States (September 2007).

[7]. Massachusetts Institute of Technology, "Carbon Capture and Sequestration Project Database," http://sequestration.mit.edu/tools/projects/index.html, accessed June 18, 2012.

[8]. Department of Energy, Energy Information Administration, Electric Power Annual 2010 (November 2011), Table 1.2, www.eia.gov/electricity/annual/pdf/table1.2.pdf. That capacity figure includes many older plants, which tend to be smaller than those being built today.

[9]. Department of Energy, National Energy Technology Laboratory, "Innovations for Existing Plants: CO_2 Emissions Control—Program Goals and Targets," www.netl.doe.gov/technologies/coalpower/ewr/co2/goals.html, which is based on Research and Development Goals for CO2 Capture Technology, DOE/NETL2009/1366 (December 2011).

[10]. FutureGen Alliance, "FutureGen 2.0" (February 24, 2011), p. 1, http://futuregenalliance.org/pdf/FutureGenFacts.pdf.

[11]. Joint Committee on Taxation, Estimates of Federal Tax Expenditures for Fiscal Years 2011–2015, JCS-1-12 (January 17, 2012), p. 34.

[12]. For a similar analysis of engineering studies, see Peter Folger, Carbon Capture: A Technology Assessment, CRS Report for Congress R41325 (Congressional Research Service, July 2010), pp. 17–19.

[13]. See Michael Hamilton, Howard J. Herzog, and John E. Parsons, "Cost and U.S. Public Policy for New Coal Power Plants with Carbon Capture and Sequestration," Energy Procedia, vol. 1, no. 1 (February 2009), p. 4489. For a discussion of the difference between subcritical and supercritical pulverized coal, see Massachusetts Institute of Technology, The Future of Coal: Options for a Carbon-Constrained World (Cambridge, Mass.: MIT, 2007), pp. 17–22, http://web.mit.edu/coal/.

[14]. A power plant's electricity-generating capacity is measured in kilowatts; the electrical power produced by that capacity is measured in megawatt-hours. During a full hour of operation, 1 kilowatt of capacity produces 1 kilowatt-hour of electricity. A megawatt-hour is 1,000 kilowatt-hours. The average cost of electricity over a plant's lifetime, sometimes called the levelized cost, takes into account such factors as the cost of building the plant, debt service, the return on equity investment, taxes, fuel costs, operating expenses, and the plant's electricity-generating capacity. For a discussion of methods used to estimate levelized costs, see Congressional Budget Office, "The Methodology Behind the Levelized Cost Analysis" (supplemental information for Nuclear Power's Role in Generating Electricity, May 2008).

[15]. Edward Rubin and others, "Use of Experience Curves to Estimate the Future Cost of Power Plants with CO2 Capture," International Journal of Greenhouse Gas Control, vol. 1, no. 2 (2007), pp. 188–197.

[16]. Peter Folger, Carbon Capture: A Technology Assessment, CRS Report for Congress R41325 (Congressional Research Service, July 2010), pp. 76–88. The decline in costs refers only to costs for the components related to reducing sulfur emissions, which in many cases fall much more rapidly than costs for the entire plant.

[17]. Edward Rubin and others, "Use of Experience Curves to Estimate the Future Cost of Power Plants with CO2 Capture," International Journal of Greenhouse Gas Control, vol. 1, no. 2 (2007), pp. 188–190.

[18]. Based on data from Department of Energy, Energy Information Administration, Electric Power Annual 2010 (November 2011), Table 1.2, www.eia.gov/electricity/ annual/pdf/table1.2.pdf.

[19]. See Table H4 in Department of Energy, Energy Information Administration, International Energy Outlook 2010, DOE/EIA0484 (July 27, 2010), www.eia.gov/forecasts/archive/ieo10/index.html.

[20]. Department of Energy, Energy Information Administration, Annual Energy Outlook 2012, DOE/EIA-0383(2012) (June 2012), Table A9, www.eia.gov/forecasts/aeo/.

[21]. That differential falls within the commonly estimated range of damage caused by CO_2 emissions. One interagency federal analysis estimated that the average social costs of the damage from climate change associated with CO_2 emissions range from \$5 to \$37 per metric ton in 2010 dollars. See Interagency Working Group on Social Cost of Carbon, Technical Support Document: Social Cost of Carbon for Regulatory Impact Analysis Under Executive Order 12866 (February 2010), pp. 1 and 33, www.epa.gov/otaq/climate/regulations/scc-tsd.pdf.

[22]. Sergey Paltsev and others, The Cost of Climate Policy in the United States, MIT Joint Program on the Science and Policy of Global Change, Report 173 (April 2009), http://globalchange.mit.edu/files/document/MITJPSPGC_Rpt173.pdf.

[23]. Department of Energy, Energy Information Administration, International Energy Outlook 2011, DOE/EIA-0484(2011) (September 2011), Table F4, p. 254, www.eia.gov/forecasts/ieo/pdf/0484(2011).pdf.

[24]. See Global CCS Institute, The Global Status of CCS: 2011 (Canberra, Australia, 2011), pp. 15–25.

[25]. For more information, see Peter Folger, Carbon Capture: A Technology Assessment, CRS Report for Congress R41325 (Congressional Research Service, July 2010), pp. 76–88.

[26]. For more information, see Congressional Budget Office, Federal Climate Change Programs: Funding History and Policy Issues (March 2010).

End Notes for Appendix

[1]. A power plant's electricity-generating capacity is generally measured in megawatts (1 megawatt equals 1,000 kilowatts), and the electrical power generated by that capacity is measured in megawatt-hours. During a full hour of operation, 1 megawatt of capacity produces 1 megawatt-hour of electricity, which, according to statistics compiled by the Department of Energy's Energy Information Administration, can power roughly 800 average households. Total plant costs do not include preproduction costs, inventory capital costs, financing costs, or costs to pay off debt while the plant is being built. CBO used total plant costs rather than a more inclusive measure because many of the studies used the total costs metric.

[2]. The \$10 cost per short ton for storage and transportation is taken from Jay Apt and others, Incentives for Near-Term Carbon Dioxide Geological Sequestration (Carnegie Mellon University, Carnegie Mellon Electricity Industry Center, October 2007), p. 4.

INDEX